NUMERICAL CONTROL

NUMERICAL CONTROL:
PRACTICE AND APPLICATION

William J. Patton

Red River Community College
Manitoba, Canada

RESTON PUBLISHING COMPANY, INC., Reston, Virginia

© 1972 by
RESTON PUBLISHING COMPANY, INC.
Reston, Virginia.

All rights reserved. No part of this book may be reproduced in any way, or by any means, without permission in writing from the publisher.

10 9 8 7 6 5 4 3 2 1

ISBN: 0-87909-564-4

Library of Congress Catalog Number: 72-85266
Printed in the United States of America

PREFACE

Courses in numerical control are found in virtually all junior colleges, colleges and technical institutes in the United States and Canada. Generally such courses have been well conceived and planned, and the author has been impressed with the quality of work put out by many such institutes and by individual students. Therefore an elementary book on the subject of numerical control would seem to serve no very useful purpose. On the other hand, this book cannot be described as advanced. But it should advance the student as far as a book will take him—experience must carry him further.

The book discusses most of the numerically-controlled machines found in the above-mentioned schools. The discussions and applications in the book, however, go beyond machining practice and should be of interest and assistance to drafting, design, and mathematics students and instructors. The author does not care to reinforce the all-too-common assumption that numerical control is a kind of machining activity.

To present elegantly and intelligently a course in APT computer programming is surprisingly difficult. There is such a wealth of APT routines, definitions, and methods. Such a course, too, must indicate to the student that many seemingly perfect APT programs are treated with contempt by the computer. The chapters on APT select only the most useful definitions and routines, and also those of greatest interest to students. A rather concentrated explanation of three languages suited to smaller computers is also included: QUICKPOINT, UNIAPT, AND AUTOSPOT. But as a subject of study, APT is the most popular by a wide margin.

The book uses repeatedly as examples a set of part prints which are grouped in the middle of the book. These are worked by a variety of methods and may serve also as a source of other ideas for numerical control work.

The intent of the book is to stimulate students to use their imaginations and to apply the methods of numerical control to a wide range of activities. To that end, the final chapter presents in some detail the method of sculpturing an artificial limb on a simple two-axis mill. It is complete with data for an actual limb. Numerical control components offer interesting projects for students of machine design or of fluid power courses also. A few of these are suggested in this book. A pneumatic position-control device of interest to students is given in *Hydraulics & Pneumatics* magazine, October 1971, page 83: Digital All-Pneumatic Positioning with Punched Cards or Tape by D. Pittwood of IBM Corporation.

The author's experience recommends the maximum of self study of numerical control by the student and the minimum of lecturing by the instructor. Students can teach each other through informal discussion and argument, and like it that way. Should any student or instructor require assistance or computer programs, he may direct his inquiries to the author through the publisher.

William J. Patton

CONTENTS

PART 1 BASIC APPROACHES TO NUMERICAL CONTROL   *1*

1 THE PRINCIPLES OF NUMERICAL CONTROL   *3*

    1.1 Automation, 3
    1.2 General Procedures for Numerical Control, 8
    1.3 Machine Axes, 10
    1.4 Positioning and Contouring, 14
    1.5 Direct Numerical Control, 15
    Questions, 16

2 SOFTWARE: CARDS AND TAPE   *19*

    2.1 Magnetic Tape, 19
    2.2 Punched Cards, 20
    2.3 Computers, 21
    2.4 Paper Tape, 24
    2.5 The Tape Code for Punched Tape, 27
    2.6 Tape Formats, 31
    2.7 Word Address Format, 31
    2.8 Sequential Format, 35
    2.9 Fixed Block Format, 35
    2.10 Miscellaneous Functions, 36
    Questions, 37

PART 2 MANUAL PROGRAMMING METHODS 39

3 PROGRAMMING IN TWO AXES 41

3.1 Basic Knowledge for Numerical Control Programming, 41
3.2 The Vertical-Spindle Cintimatic, 42
3.3 Setting Up the Work Piece, 44
3.4 The Information Block, 45
3.5 Sequence Number, 46
3.6 Miscellaneous Functions, 47
3.7 A Look at the Part Program, 48
3.8 Manual Control of Z-Axis Movements, 49
3.9 The Preparatory Functions, 50
3.10 Programming Considerations, 52
3.11 Tab Sequential Programming, 53
Questions, 54

4 CONTOURING OPERATIONS WITH A POSITIONING MILL 57

4.1 Pocket Milling with a Positioning Mill, 57
4.2 Angle Cuts, 59
4.3 The General Case of the Circle, 60
4.4 Programming the O-Ring Groove, 62
4.5 Preparing and Checking the Tape, 63
Questions, 64

5 FIXED BLOCK PROGRAMMING 67

5.1 The Fixed Block Format, 67
5.2 The Moog Hydra-Point 83-1000 MC, 68
5.3 The Information Block, 69
5.4 A Programming Example, 72
Questions, 72

6 OPERATIONS IN THREE AXES 75

6.1 The Three-Axis Machine, 75
6.2 Programming the Milwaukee-Matic Model II, 78

Contents

6 OPERATIONS IN THREE AXES—Continued

    6.3 Coordinate Axes of the Model II, 81
    6.4 The Part Program for Positioning Control, 82
    6.5 The Program Sheet, 85
        Questions, 91

7 BASIC PRINCIPLES OF CONTOURING   **93**

    7.1 Contouring Computations, 93
    7.2 Contouring by Linear Interpolation, 94
    7.3 Circular Interpolation, 99
    7.4 Parabolic Interpolation, 101
    7.5 Curve-Fitting, 102
    7.6 Tool Offset Calculations, 106
        Questions, 114
    7.7 Tape Reading Speed and Block Process Time, 115
    7.8 Standard Address Codes for Numerical Control, 116
    7.9 Preparatory Functions for Contouring Machines, 117
    7.10 Contouring Machines, 117
    7.11 The Gorton Tapemaster 2-30 Contouring Mill, 119
    7.12 Feed Rate Number, 120
    7.13 A Programming Example for the Gorton Tapemaster, 122
    7.14 Slope and Arc Cuts on the Model II Milwaukee-Matic, 122
    7.15 The Cintimatic Family of Numerically Controlled Machines, 124
    7.16 Programming the Cintimatics, 126
    7.17 Z-Axis Movements for the Turret Drill, 129
    7.18 Unit Operations with the Turret Drill, 130

8 THE NUMERICALLY CONTROLLED LATHE   **135**

    8.1 Conventional Practice for N/C Lathes, 135
    8.2 Dimensioning the Tool Path, 138

8 THE NUMERICALLY CONTROLLED LATHE—Continued

8.3 Dimensioning Tool Displacements for Taper Cuts, 141
8.4 The Monarch Turn/Center 75 Two-Axis Lathe, 142
8.5 Example 1: Straight Turning, 150
8.6 The Program, 151
8.7 Arc Offsets, 155
8.8 Direct Feed Rate, 155
8.9 Example 2: Programming A Seating Pin, 156
8.10 Thread Cutting, 156
8.11 A Thread Cutting Example, 159
8.12 Practice Examples for N/C Lathe Programming, 160

9 THE HARDWARE OF NUMERICAL CONTROL 165

9.1 Reading the Tape, 165
9.2 Processing the Tape Information, 167
9.3 Linear and Circular Interpolation, 168
9.4 Open and Closed-Loop Control, 169
9.5 Programming the Slo-Syn Installation, 171
9.6 Positioning Programs for the Slo-Syn, 172
9.7 Contour Milling with the Slo-Syn, 174
9.8 Contamination of Hydraulic Oil, 177
9.9 Tooling, 178
9.10 Inspection, 180
9.11 Drafting, 181
9.12 Flame Cutting and Welding, 182
9.13 Punching, 182
9.14 Machine Specifications, 183
9.15 Adaptive Control and the Information System, 183
Questions, 185

APPLICATIONS OF NUMERICAL CONTROL 187

Schedule of Part Prints for Practice in Numerical Control

Contents   xi

10   APT: THE BLUE-COLLAR COMPUTER LANGUAGE   *197*

    10.1 APT and Numerical Control, 197
    10.2 Description of the O-Ring Groove, 199
    10.3 The APT Programming Sequence, 199
    10.4 A Survey of the Tool Movements, 203
    10.5 The Program Printout (CLPRNT), 204
    10.6 Preparing the APT Manuscript, 206
    10.7 General Information Statements, 209
    10.8 The Structure of APT Geometric Statements, 211
    10.9 The APT Processor System, 213
            Questions, 214

11   APT GEOMETRIC DEFINITIONS   *215*

    11.1 General Concepts for Surface Definitions, 217
    11.2 POINT Definitions, 217
    11.3 Line Definitions, 217
    11.4 Plane Definitions, 221
    11.5 Circle Definitions, 221
    11.6 CYLNDR Definitions, 224
    11.7 A Programming Suggestion, 226
            APT Practice *Questions,* 226

12   CONTROLLING THE TOOL MOVEMENTS   *229*

    12.1 First Principles, 229
    12.2 GOTO/ and GODLTA/, 230
    12.3 ZSURF/, 231
    12.4 Tool Movement Control in Contouring, 232
    12.5 The Start-Up Statement, 234
    12.6 Go Movement Statements, 235
    12.7 Special Cases in Tool Motion, 237
    12.8 Program Printouts, 241
    12.9 Some Observations for Programming APT, 241
    12.10 Nested Definitions, 242
            Questions, 243

13 THE POSTPROCESSOR 249

13.1 General Concepts, 249
13.2 The MACHIN/ Statement, 251
13.3 SPINDL/, 251
13.4 COOLNT/, 252
13.5 FEDRAT/, 252
13.6 TRANS/ or ORIGIN/, 252
13.7 SEQNO/, 253
13.8 MCHTOL/, 253
13.9 STOP, 253
13.10 END, 253
13.11 LEADER/, 254
13.12 AUXFUN/, 254
13.13 CLEARP/ and RETRCT, 254
13.14 CYLE, 254
13.15 PREFUN/, 255
13.16 DELAY/T, 255
13.17 LOADTL/, 255
13.18 RAPID, 255
13.19 SELCTL, 255
13.20 ROTABL, 255
13.21 PLUNGE/, 256
13.22 FLAME/, 256
13.23 DRAFT/, 256
13.24 POSTPROCESSING WITHOUT A MANUAL, 257

14 VERSATILITY IN APT 259

14.1 Moment of Inertia and Surface Area, 259
14.2 The MARCO Feature, 261
14.3 The POCKET/ Routine, 263
14.4 PATTERN/ for Hold Patterns, 266

15 SMALL COMPUTERS AND SPECIAL LANGUAGES 271

15.1 ADAPT, 271
15.2 UNIAPT, 272
15.3 Quickpoint, 273
15.4 Contouring with Quickpoint, 277

Contents

15 SMALL COMPUTERS AND SPECIAL LANGUAGES—Continued

15.5 A Quickpoint Programming Example, 278
15.6 AUTOSPOT, 279
15.7 AUTOSPOT Geometric Statements, 279
15.8 Types of AUTOSPOT Statements, 282
15.9 Specification Statements, 283
15.10 Datum Points, 284
15.11 Pattern Definitions, 285
15.12 Machining Statement, 287
15.13 Pattern Manipulations, 291
   Questions, 292
15.14 Postprocessing of AUTOSPOT Programs, 292
15.15 Tool Statements, 293
   Questions, 294
15.16 Milling Operations, 295
   Questions, 300

16 THE PROGRAMMER AS SCULPTOR *301*

16.1 QADRIC, 302
16.2 GCONIC, 303
16.3 A QADRIC Example, 305
16.4 TABCYL, 319
16.5 A TABCYL Application, 320
16.6 Analyzing the TABCYL, 323
   Questions, 330

17 CREATIVITY IN NUMERICAL CONTROL *335*

17.1 A Technical Frontier: Artificial Limbs, 334
17.2 Sculpturing a BK Model, 335
   Questions, 337

ANSWERS TO QUESTIONS IN THIS BOOK *341*

INDEX *351*

PART 1

BASIC APPROACHES TO NUMERICAL CONTROL

1

THE PRINCIPLES OF NUMERICAL CONTROL

1.1. AUTOMATION AND INFORMATION PROGRAMMING

A casual review of the careers open to men and women today suggests that most jobs are largely concerned with the processing of information. Information is sought, then recorded, compiled, and processed, and finally decisions are made and action is taken on the "facts," that is, information, so processed. The initial step in a medical treatment is the search for the pertinent medical facts; the processing of this information is called a diagnosis. Similarly, a lawyer first obtains information in order to initiate a lawsuit. A draftsman processes information into graphics, whereas an engineer analyzes physical information about hardware. Even a customer in a store first seeks information on price, style, and quality before he buys the goods.

If you wish a machinist to make a part on a machine tool, he first asks for a sketch or a blueprint, which is a compilation of the information into a form readily communicable to the machinist. He then reprocesses the information into his special technology: a certain material of a certain bar size, hardness, machinability, tolerance, and surface finish, a certain machine speed and feed, a certain cutting coolant, a certain cutting tool material and shape. The machinist must be a thinker before he undertakes to sculpture metals, and the best machinist is he who can process the widest range of information within his technology.

The technology of numerical control has invented no new processes. Numerical control is simply a relatively new method of organizing the information required for a process and a new method of inserting this information into the process. Since numerical control is a type of automation, its significance for the present and future comes to view in a general discussion of automation.

Four types of automation are at present in use, and there would seem to be little possibility of other types being developed.

1. *Process control.* Process control is largely concerned with chemical and physical processes that involve flow of fluids, pressure, and temperature. Common examples would be the automatic control of heating boilers, climate control of buildings, or control of oil refinery processes. This is a type of automation that at present does not require the methods of numerical control, and need not be further discussed.
2. *Electronic data processing* (EDP). EDP refers to the processing of data by means of computers, usually of the electronic-mechanical type, usually referred to as electronic. Such automatic processing of information is frequently required in preparing complex numerical control programs, and therefore is a necessary subject of this book.
3. *Fixed automation.* Also called Detroit automation. Though it is not in general use for the assembly of automobiles, it is commonly in use to produce automobile components. This type of automation is basically suited to long production runs.
4. *Numerical control* (n/c). A flexible method of automation applicable to short production runs or even to a "one-off" part.

Fixed automation is suited to the production of any part required in very large quantities, such as automobile engine blocks, spark plugs, telephone parts, instamatic camera cartridges, newspapers, soft drinks, and electrical switches and thermostats. Let us use an automatic lathe as an example of fixed automation. By means of cams and other devices, a variety of cutting tools on the automatic lathe is sequenced to approach and machine the part in a fixed sequence from which there can be no variation. The resulting production is some hundreds or thousands of identical production parts, identical because the same sequence of tool movements was used with each one. The automatic lathe can be converted to the production of a different part, but only after resetting of the tools and cams to provide the required new sequence of movements. The work of setting up such a machine for a production run requires half a day to a couple of days; more complex automatic machines may require weeks to set up.

Such complex set-up procedures are bypassed in numerical control. Since the procedures of numerical control are the extended subject of this book, we may for the moment oversimplify the case by saying that in

(a)

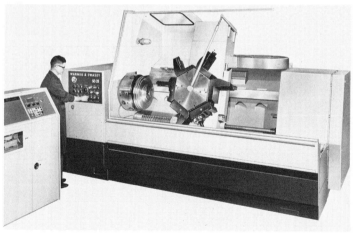

(b)

Fig. 1. Warner and Swasey lathes, (a) single-spindle automatic lathe with six-station tool turret above the lathe spindle, no cutting tools are mounted in the turret; (b) numerically controlled lathe with six-station tool turret (note the simpler tooling configuration as compared with the automatic lathe). On the left is the machine control unit which contains the tape reader and the electronic computing and control circuitry.

numerical control operations the sequencing of movements is controlled by instructions given to the machine by a code of holes punched into paper tape. Therefore, to change the part to be produced, we simply insert the required punched paper tape in the tape reader of the machine, and the machine is ready. The set-up procedure, therefore, lies in programming and punching the tape, and, in contrast to fixed automation, this set-up work is done *away* from the machine—a very significant advantage, since the machine can be working on one part while the next part is being programmed. By contrast, the automatic lathe is shut down for set-up operations.

Computer programs and languages, punched cards, magnetic tape, paper tape, and other such information-processing items are referred to as *software*. In contrast to the methods of fixed automation, numerical control requires the use of software for its information processing. (Someone will argue that in order to set up the automatic lathe, software is also required, a blueprint of the part at least.)

What, then, is the essential difference between fixed automation and numerical control? In fixed automation the program or sequence of operations is built into the *hardware,* whereas in numerical control the program is in the *software*.

The separation of the information programming from the operational hardware, as in numerical control, is perhaps one of the very great inventions of this century. The employment of this principle is certain to be extended by sociological trends. Fixed automation reduced manufacturing costs to minimum levels, making possible what is termed "the affluent society." The penalty of such low costs, however, is that the consumer must buy the same product as everyone else. Under fixed automation, there is little scope for variation, because that is too costly. Variations tailored to the wishes of individual customers are a minor inconvenience in the methods of numerical control; a small section of the punched tape must be reworked. Certainly the consumer market shows a strong trend to greater diversity in product "options," and numerical control makes such options feasible. We shall in the future be less concerned with engineering "efficiency," since to some degree it is a dehumanizing influence. Instead, we shall want what we want when we want it. Numerical control belongs in such a world.

The study of numerical control, therefore, is largely the study of information programming routines. These are somewhat similar to the routines used in computer programming. This leaves a first impression that there is nothing to know about n/c except a standard procedure—a procedure that can be learned in two hours to two weeks, depending on the complexity of the n/c machine. The tape goes into the machine control unit, and the part is automatically produced.

The author remembers a sales presentation for an n/c drilling machine. The speaker rather enthusiastically predicted that the machinist trade would become obsolete very quickly, since n/c operations do not require the machinist's skill. Before we consign the departing machinist to history and replace him with a programmer and an unattended n/c machine tool, perhaps we should look a little more closely at the operations that n/c programs call for. Certainly the machinist's manual skills are not needed when the n/c machine has them, but we need the machinist's knowledge and experience. These are required if the proper speed and feed are to be included in the n/c program. The virtues of n/c evaporate if only broken drill bits are produced. When more complex work is programmed, even greater demands are made on the machinist's knowledge. Nor is it entirely true that the skill required in producing the part by n/c lies entirely in the programming. The program does not usually affix the work piece to the machine table, nor locate it in the proper place. Further, the programmer cannot foresee all possibilities. If the part is an iron casting, the programmer may set the speed and feed at a usual figure for soft gray iron. Now suppose we use an office girl for the n/c machine operator, on the theory that neither skill nor knowledge is needed. The casting may have picked up some molding sand. The office girl does not see such sand inclusions as pertinent information. If we replace her with an experienced machinist, he knows what molding sand can do to his cutting tool, and perhaps he will override the tape instructions by switching the machine to manual operations and manually finish the piece. He can correct such difficulties as tool chatter. Or the programmer may make a mistake that an experienced machinist will catch. Therefore, we must recognize that a machining operation remains a machining operation regardless of the method of putting the information into the process.

Certainly, programming is the major concern in n/c production, but this book will try to show some of the more practical difficulties in the management of the hardware and the n/c operations. Students of n/c programming need not run every program through a numerically controlled machine, but a good programming course requires that as much of the trainee's time as possible must be spent watching the execution of his programs. Otherwise, he misses too much. Programming, even computer programming for numerical control, is not especially difficult, but neither is it irresponsibly easy.

Most n/c applications are in machining operations, chiefly drilling and milling. The methods of numerical control, however, have had a wide application. They are used in drafting, painting, sandblasting, welding, flame-cutting, punching, material-handling, adhesive bonding, assembly operations on electronic equipment, and others.

There are applications for which numerical control is unsuited. Since it is a method of information processing, its utility increases with the amount of information to be processed, or rather with the *rate* at which information must be processed. If a machinist must turn 1-in. bars 12 in. long down to $\frac{7}{8}$-in. diameter, the amount of information processing is minimal, and the complex tape-reading and control apparatus of the n/c lathe cannot be justified. But if the machinist must produce the complex contours of an artificial limb on a lathe, there is no alternative to numerical control. The extent of its application to machining operations simply indicates that more information is processed in machine shop work than in other operations.

Numerical control may also be justified if it is desired to process information more rapidly, or if information is certain to be altered. It may be necessary to produce a prototype part to be tested and modified after the tests. Numerical control is used under these circumstances, since the modifications necessitate only minor changes to the control tape. Such changes may be very costly with other methods, especially if tooling or fixtures must be modified.

In general, these circumstances would seem to require numerical control:

1. Small production quantities, generally below a thousand, or as little as one.
2. Complex and difficult operations requiring a high degree of skill, such as the example of an artificial limb.
3. Difficult tolerances.
4. Components which otherwise result in high percentages of scrap and rework, especially if the material is expensive.
5. Replacement parts.
6. Prototype work, or parts subject to modification.
7. Parts which otherwise require expensive tooling.
8. Production subject to tight delivery schedules, since numerical control does not require the time and expense of designing and building expensive fixtures and other tooling.

1.2. GENERAL PROCEDURES FOR NUMERICAL CONTROL

Programming of a numerical control operation is executed either manually or with the assistance of a computer.

Most programming is done manually. The flow chart for a manual program is given in Fig. 2. The programmer writes out the machine instructions in tabulated blocks of information on a program manuscript. These instructions are then punched into the control tape on a suitable tape punch. The control tape is inserted into the tape reader of the numerical control machine. The machine reads a block of information at a time and executes the instructions in the block, after which the next block is read. If the operation is the drilling of holes, the location of each hole to be drilled is given in a separate block of information. The machine

Figure 2

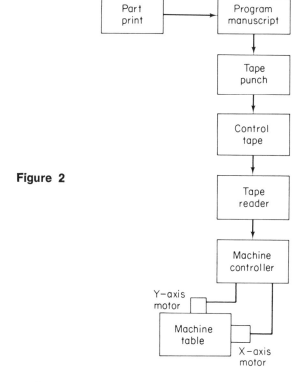

reads the location of the next block, positions the work piece for the drill, then drills, then reads the next block, and so on.

Figure 3 is the flow chart for a computer-executed n/c program. Computer assistance is required, for example, in the numerically controlled drafting of a ship's contours, or the teeth of a gear.

The part programming in this case is quite different from manual programming. The tape reader of the n/c machine must still read the blocks of information that its tape reader and electronic logic system are designed to receive, but the computer, not the programmer, prepares the information blocks for the tape. The program manuscript does not contain such blocks. Instead, it consists of computer language statements suited to numerical control. Each statement is punched into a computer card, and the card deck is processed by the computer to give a geometrical solution for the part geometry. This is a general solution, suitable for any numerical control machine. That is, the computer solution could be applied to an n/c drafting machine to draw the part or to an n/c machine to make the part, or to an n/c inspection machine to inspect the part. Actually to make the part, or draw it, or inspect it, the general computer solution must be modified to adapt it to the special requirements of the actual n/c machine to be employed. Every n/c machine must have its blocks of information arranged in a special format, for example. The general solu-

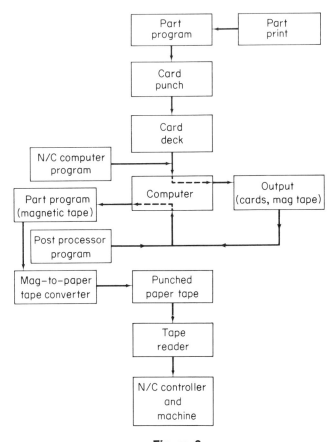

Figure 3

tion is therefore *postprocessed* (reprocessed). From the second computer run a suitable punched tape is prepared for insertion into the tape reader of the machine.

Later in this book the programming of an elliptical groove is extensively discussed. A full day is required to program the groove by manual methods. Perhaps 20 minutes are required to prepare a computer program for the same job, and the computer cost is only a few dollars. Many contouring operations are virtually impossible to execute by manual programming methods.

1.3. MACHINE AXES

The machine instructions punched into the tape are of two kinds:

1. Geometrical machine movements, the basic part of the program.

2. Auxiliary operations, such as turning on coolant, reversing the spindle, changing the tool, etc.

Machine movements, like all movements, are of two kinds: translation and rotation. Translation is a displacement in any of three directions, usually designated X, Y, and Z for purposes of numerical control; similarly, rotation is possible about three possible axes mutually at right angles, one axis oriented in the X direction, one in Y, and one in Z.

Any movement, translation or rotation, which is under the control of tape-coded instructions is referred to as an *axis*. Most n/c drilling machines are two-axis machines, controlling table movement in the X and Y direction in order to position the drill spindle. The third or Z movement in such a machine is fully or partially under manual control. If the movement of the drill spindle in and out is also under tape control, then the drill is a three-axis machine. A four-axis machine would control movements in X, Y, and Z, and in addition would have a rotating work table also under tape control.

Most n/c machines use what is referred to as an absolute positioning geometry, which means that all measurements are taken from a fixed origin with coordinates $X = 0$, $Y = 0$, $Z = 0$. This origin has different locations on different machines. Often the origin is a point on the work table surface. Most numerically controlled lathes do not use an absolute positioning system with fixed origin, but an *incremental* system. In an incremental system, all measurements for the next position are made from the last previous position. The two positioning methods are illustrated in Fig. 4.

The industry standard for axis and motion identification is Electronic Industries Association (EIA) Standards Proposal No. 731. The Z axis of motion is the axis through the center of the spindle. Since machine tools are built in both horizontal and vertical spindle types, the Z axis can be either horizontal or vertical. Since most lathes are horizontal-spindle machines, the Z axis for such a lathe will be horizontal. See Fig. 5. The positive Z direction is in the direction from the work-holding means toward the tool-holding means, a statement that may be confusing. It means that $+Z$ is toward the spindle from the work table, and $-Z$ is away from the spindle, if the work table holds the work and the spindle holds the tool. For a lathe, the headstock spindle rotates the work, and the toolholder is mounted on a tape-controlled slide; therefore, for a lathe the $-Z$ direction is toward the spindle from the slide.

The X axis is horizontal, and for most machines such as milling and drilling machines is the long axis of the work table (see Fig. 6). If Z is vertical, $+X$ is to the right, looking from the front of the machine to the back. If Z is horizontal, $+X$ is to the right looking from the spindle to the work piece.

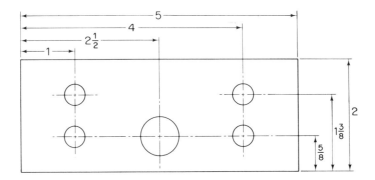

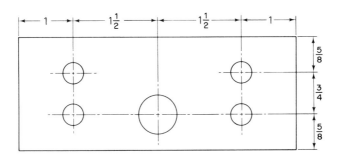

Figure 4

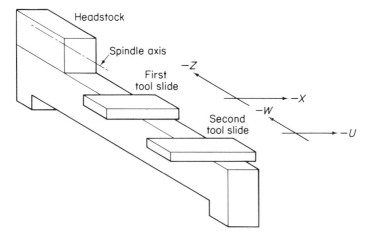

Figure 5

For vertical spindle mills and drills, the operator standing in front of the machine is looking into the machine in the $+Y$ direction.

Positioning and Contouring

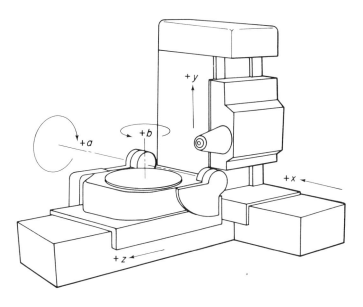

Figure 6

The rotary motions are designated A, B, and C. A corresponds to rotation about the X axis, B to Y, C to Z. Positive A, B, C are such as to advance a right-handed screw in the $+X$, $+Y$, or $+Z$ direction.

A general acquaintance with these axis specifications is sufficient. The specific axis system of the machines you must program must be more thoroughly understood.

The cartesian coordinate system is standard in n/c dimensioning. The quadrants are numbered counterclockwise in accord with standard mathematical convention. See Fig. 7.

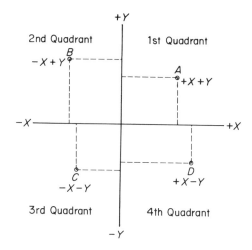

Figure 7

When one is punching coordinate dimensions into the tape, the decimal point is omitted, or, more unequivocally, must not be used. Most numerically controlled machines can position to three decimal places. Thus if $X = 3.125$, this dimension is punched 03125.

1.4. POSITIONING AND CONTOURING

As may be seen in Fig. 1(b), a numerically controlled machine is packaged in two parts: the processing machine and the machine control unit. The machine control unit contains the tape reader and electronic circuitry that controls the positioning motors operating the axis motions.

Two types of control systems are used to operate n/c machines, numerical positioning control (NPC) and numerical contouring control (NCC). A third type has recently been introduced, direct numerical control (DNC).

The simpler type of control is positioning control. Its function is to move the machine table or machine spindle to the required position at which the operation is to be performed—drilling, tapping, boring, insertion of a component, etc. The positioning and the processing are done in sequence in positioning control. Such a system is suited to a drilling machine, for example.

Relatively slow tape readers are employed for positioning control, and reading speed would not exceed 60 characters per second. A block of information for such a machine might contain about 20 characters, letters and digits, so that a block of information would be read from the tape in 0.3 second or less. This is adequate reading speed, since operation time for the process would be much longer. The work table might require a few seconds to reach the position commanded by the tape, while drilling the hole might require as much as half a minute. Because of the relatively short tape reading time, the next block of tape information can be read as soon as the previous operation is completed.

In positioning control, since no operation is performed during positioning, the path taken to arrive at the command position need not be restricted or controlled (except perhaps to avoid a collision between fixturing and spindle). Usually the machine positions at the same velocity in both the X and Y directions, thus moving at 45 degrees to both axes until positioned in one axis, then completing the movement in the longer direction. This is shown in Fig. 8. Thus if the machine movement was from $X = 0.0, Y = 0.0$ to $X = 6.0, Y = 9.0$, the movement would be such as to reach first $X = 6.0, Y = 6.0$, then the final movement to $Y = 9.0$.

Some positioning machines are equipped with a milling spindle and feed rate control during positioning, and when thus equipped can, in addition to drilling and tapping operations, perform straight-cut milling parallel

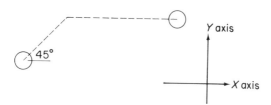

Figure 8

to either X or Y axis. Such a relatively simple machine capability is sufficient for much contouring work, and the contouring methods for it are explained in a later chapter.

A continuous path or contouring system has independent control of the speed of the X and the Y drive. Thus milling and profiling can be performed at any and all angles to the two axes. When suitably programmed, contouring machines can cut any complex curve in two or three dimensions or even sculpture. Drafting machines and flame-cutting machines require a contouring control system.

Contouring controls must process tape information at more rapid rates. In the contouring examples to be discussed in later chapters, the tool makes hundreds of small machining movements, perhaps only a few thousandths of an inch in the X or Y direction. In such operations the machining time becomes as brief as the tape-reading time. High-speed tape readers of the photoelectric type are required. The n/c machine cannot wait for the tape reading operation, and a system of tape reading called buffer storage is adopted. A block of tape information is read into the buffer storage (memory) of the control. This information block is then transferred to active storage. The n/c machine operates with the information in active storage, while simultaneously the next block of information is read from the tape and stored in the buffer storage. By this means the tape reading operation can keep up with the machine process.

Most positioning machines use absolute dimensioning from an origin. Contouring machines use incremental positioning from the last position. Since n/c lathes are equipped with contouring controls, incremental dimensioning is characteristic of the lathes discussed in Chapter 8.

1.5. DIRECT NUMERICAL CONTROL

The third type of numerical control system is direct numerical control (DNC). The economics of this system are somewhat involved, and the method is suited only to very complex n/c systems, such as multi-axis machines or systems of operating several n/c machines. In this system there is no tape reader, no tape, and no machine control unit. A computer performs the machine calculations and generates the electrical impulses to move the motors that position the work table or spindle. The basic problem in this control method is one of incompatibility in information processing; any computer, large or small, can process information at a speed far beyond the rate at which the n/c machine can use it.

However, DNC has advantages to be exploited:

1. The computer can program simultaneously many n/c machines.
2. The computer can be remotely located, even a thousand miles away.
3. The tape preparation, the tape reader, and the control unit are not needed.

It is surely reasonable to prophesy that, as computer technology evolves further, before the end of this century DNC will be used to link up all n/c machine installations in a continent-wide manufacturing network. Certainly the technology for doing so is already available. Manufacturing could then be executed in a geographic location convenient for delivery to the customer but controlled from a location convenient for administration. The computer used for n/c control could, of course, process all job control, costing, invoicing, and other documentation. The trend to more flexible manufacturing output mentioned earlier would add further impetus to such a development. If these predictions are credible, then machinists are going to learn computer programming, and computer programmers are going to study manufacturing operations, and the difference between white-collar and blue-collar operations should disappear.

Twenty years ago, about the time when the author graduated from engineering college, we would have said that a marriage between computers and manufacturing was incongruous. At that time computers were used only for scientific calculations. In 1952 the author urged his university to begin the study of computers and their applications. The recommendation was denied at that time as not useful. Twenty years is a whole era in technology, and in the thinking of men.

This book, then, is about both computers and production operations, both hardware and software.

Questions

1. Suppose the lathes, drills, mills, and grinders in your shop are all n/c equipped. Discuss and sketch the standard axis notation for each machine. What axis designations would you apply to a radial drill?
2. Figure 1-1 is a plate to be drilled with holes as shown, absolute positioning being used.
    a. The origin of coordinates is at the center of the plate as shown.

Dimension the holes for n/c programming with five digits, three decimal places, and plus or minus sign. For example, hole no. 1 is dimensioned —X03000—Y02000.

b. Assume the origin of coordinates to be 10 inches to the left of, and 3 inches below, the lower left-hand corner of the part. All hole positions will now be in the first quadrant. Dimension the holes for n/c programming. Since all dimensions are now positive, it is usual to omit the plus sign.

Note. In n/c programming there cannot be errors. Regardless of the simplicity of this little exercise, check your answers. If you are not absolutely certain that they are correct, you are wasting your time learning numerical control. The answer to this question is *not* given at the back of the book.

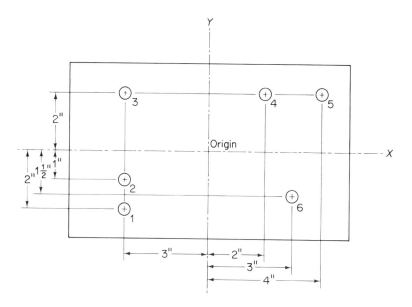

Figure 1-1

2

SOFTWARE: CARDS AND TAPE

2.1. MAGNETIC TAPE

Magnetic tape was not designed for a shop environment. It is readily harmed by dust, dirt, and the powerful magnetic fields radiated by such equipment as resistance welders and induction generators. However, magnetic tape can be used in a shop if proper handling methods are used. Magnetic tape cabinets and equipment must be kept clean, and clean gloves must be used when one is handling the tape cartridge. The standard tape cartridge appears to be adequate protection for the tape under shop conditions.

When numerical control methods were first invented, about 1958, magnetic tape was the information medium, not punched tape. It is still in use for numerical control operations. Probably magnetic tape will be increasingly used in the future for numerical control purposes, especially for contouring operations, though perhaps not for the simpler positioning systems.

The major advantage of magnetic over punched tape is its greater information storage. Punched tape can carry only 10 characters per inch; magnetic tape about 200. The punched tape for a complex n/c machining operation could be three-quarters of a mile long, but if the same data were recorded on magnetic tape, the tape length would be reduced to 200 feet.

The information on magnetic tape is recorded by magnetization of small tape areas. This information can, therefore, be erased and the tape reused. Magnetic tape can be produced by the standard equipment of a computer department, whereas punched tape is not well adapted to computer data processing because of the slow speed of tape punches. Punched tape, however, is cheaper, and a tape reader for punched tape is considerably cheaper than a magnetic tape reader.

No complete standardization of either punched or magnetic tape character codes has yet been achieved, and the lack of standardization will become an increasing problem the longer it is delayed. Manufacturing operations, commerce, technology, and information transmission are already international in scope, and standardized information media are basic to such worldwide operations.

The present approach to a standard magnetic tape for numerical control and other information operations is the Aerospace Industry Association NAS968, shown in Fig. 9. There are seven tracks as compared with eight for punched tape. The character coding in the figure is designated by 1 or 0. Each digit 1 or 0 is a small magnetized area on the tape. The number 1 represents an area magnetized in the reverse direction from that of a number 0. The magnetic recording head thus reverse-magnetizes the "1" areas and passes the "0" areas. This character code is virtually identical to the new ASCII/ISO character code for punched tape discussed later in this chapter, but does not correspond to the punched tape code heretofore universally in use in this country.

Magnetic tape is made of a strip of plastic material $\frac{1}{2}$ inch wide. A thin layer of gamma ferric oxide particles held in a binder is bonded to one face of the tape. Gamma ferric oxide is a magnetic oxide, and these are the particles magnetized by the recording head. The tape must be dimensionally stable when subjected to changes of temperature and humidity, and must have a high tensile strength. It must not be too elastic because of the requirement of dimensional stability. It is preferable that the tape should break rather than stretch.

2.2. PUNCHED CARDS

The familiar punched card is used as an input device to the computer when a computerized n/c program is necessary. Both the Univac and the IBM card are the same size, $3\frac{1}{4} \times 7\frac{3}{8} \times 0.0065$, but use different information codes. The IBM card has 80 columns across the card for punching information. This card is shown in Fig. 10. The manuscript paper used for writing the computer statements to be punched into the cards likewise has 80 columns. The key punch operator normally punches a character in the same column of the card as the column in the manuscript paper in which it appears.

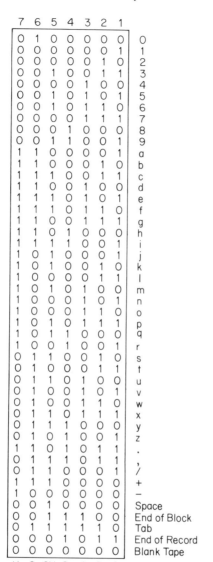

Figure 9

2.3. COMPUTERS

Except for some of the more complex APT computer routines, none of which are discussed in this book, the n/c programmer need not understand digital computer operation. But a few remarks about such computers are in order here, if only because the machine control unit of the numeri-

22 *Chap. 2 / Software: Cards and Tape*

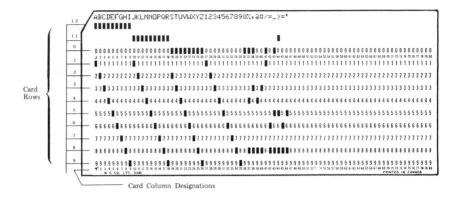

Figure 10

cally controlled installation is itself a computer on a more limited scale. The computer usually receives its data from punched cards, whereas the MCU reads from punched tape. The computer data output is a printout, a deck of cards, a magnetic tape, or other media. The MCU output is electrical signals to the drive motors that control the axis movements.

A digital computer is a complex electromechanical machine that performs lengthy logic and calculating functions. The basic machine components of a computer are shown in Fig. 11.

The *control unit* of a computer may be likened to a telephone ex-

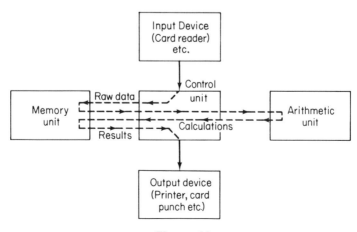

Figure 11

change; it coordinates and sequences the calculations, operating the input and output devices as required, taking data out of the memory, and so on. The *arithmetic unit* makes the calculations. Since a computer can do no other kind of mathematics than binary addition, the internal programming of such a computer must reduce elaborate mathematics to addition operations. (The analog type of computer, not used in numerical control or data processing, can perform operations in advanced mathematics.)

Data are stored in the memory or storage unit. The two words mean the same thing. Another word for data storage is *register*. A register is used for temporary storage. Since all tape-punched data entering the MCU of an n/c machine are stored only temporarily before being fed out to the machine tool, strictly speaking the MCU contains only registers. However, the words "memory," "storage," and "register" are used rather carelessly, and will be so used in this book.

Storage devices may be magnetic cores, magnetic drums, magnetic tape, magnetic disks, and other devices. It does not appear to be conventional use to consider punched cards and punched tape as computer storage devices.

Figure 12 is used as an example of computer calculation for computerized programming. The cutting tool or drafting pen must move along

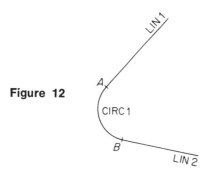

Figure 12

LIN1, around CIR1, and over LIN2. The points of tangency A and B are rarely dimensioned on part drawings and, therefore, must be calculated. A set of statements containing what is known about LIN1, CIR1, and LIN2 is punched into cards, and the card information is read into the computer. The card data are sent to the memory unit of the computer. The control unit draws information from memory and supplies it to the arithmetic unit. Intermediate solutions from the arithmetic unit are restored in the memory and drawn upon again to the arithmetic unit for further calculation. The final solutions are stored in the memory unit, then mustered out to the output device, which perhaps tabulates them on a printout, or may record them on magnetic tape for final conversion to a paper tape.

2.4. PAPER TAPE

Punched tape is the usual medium for inserting the information into a numerically controlled operation. Paper tape is convenient for shop operations and equipment, is reasonably compact, and does not require expensive equipment for tape punching or tape reading.

Paper tape can tear, especially if there are lines of many holes across it as occurs when the character "delete" is punched into the tape. Mylar tape cannot be torn, but its high shear strength produces greater wear on tape punches.

Paper tape comes in rolls about 8 inches in diameter, costing about a dollar each. One roll can record about 100,000 characters. Various colors are available so that, if desired, color coding is possible.

Tape for numerical control is 1.000 ± 0.003 in. wide, and the punched hole positions are spaced 0.100 in. each way. N/c machines use eight-track tape, which means that there are eight rows of holes across the width of the tape. Other tape systems may be five, six, or seven-track. Drive sprockets move the tape through tape punches and tape readers by engaging a row of sprocket holes punched in addition to the eight channels of information holes. Tape thickness is 0.004 ± 0.0003 in. Specified dimensions of the tape are shown in Fig. 13, as set out in the pertinent standards of the Electronics Industries Association.

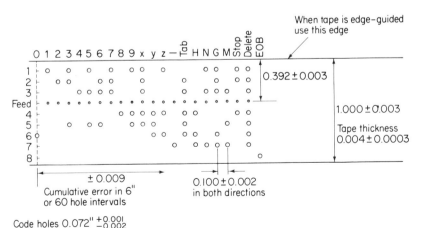

Figure 13

Paper Tape 25

Since holes are punched every 0.100 in. along the length of the tape, one inch of tape records 10 characters. However, a long leader and a long trailer are needed at the start and finish of a tape program to allow for ease of inserting the tape in the n/c tape reader, and these lengths of tape, of course, convey no information.

Short tape can be punched on manual types of punches. One type of manual punch has a handle that is turned to dial the character required, the handle being used also for the punching operation. With a very little experience the operator can punch about 60 characters a minute on such a punch.

Most tapes are prepared either by computer or a suitable automatic typewriter. The typewriters most commonly used are the Dura and the Flexowriter. The latter is discussed here, and is illustrated in Fig. 14.

On the Flexowriter, the tape punch is located behind the tape reader at the left-hand side of the keyboard. An n/c program (or letter or other document) may be typed without tape punching. But if an n/c tape is to be prepared, paper tape is threaded through the tape punch and a tape is punched character by character as the keys are struck; the in-

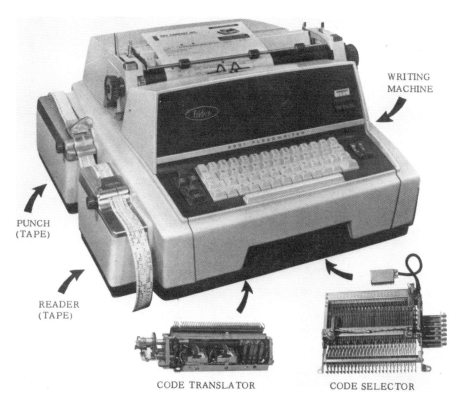

Fig. 14. The Friden Flexowriter.

formation punched into the tape is at the same time typed on the sheet of paper in the carriage of the typewriter.

After the paper tape program is punched, a second copy of it must be punched, since the first tape could be damaged or misplaced. To reproduce a tape, the original is inserted into the tape reader and a blank tape is inserted into the tape punch. As the second tape is copied from the first, the typewriter can at the same time retype the manuscript information. This second copy of the manuscript should be compared with the original in order to check for errors. The tape reader is also used to operate the typewriter automatically from punched tape. The Flexowriter, therefore, is also an n/c machine.

Figure 15 shows the switches at the left-hand side of the keyboard.

Fig. 15. Manual control switches on the Flexowriter.

Depression of the TAPE FEED switch will cause the tape automatically to feed through the punch. The START READ switch starts the tape reader; STOP READ shuts off the reader. The figure shows an unidentified switch. It is usual on Flexowriters adapted to numerical control to convert this to a DELETE key for correcting errors. This DELETE key substitutes for that all-important tool, the eraser, when the wrong code is punched into the tape.

Most typing errors are known immediately as they are made. Suppose that the position X03232 was typed as X03323, and the error was immediately noted. The tape is backed up so that the first error 3 (instead of 2) is under the punch. The DELETE key is pressed. This punches a row of holes across the tape. Similarly, the following 23 is deleted by rows of holes. The correct characters 232 are next typed and punched. The typed sheet now reads X03323232. But when the finished tape is inserted in the tape reader, the erroneous characters, which were deleted, are not typed out on the manuscript sheet, and similarly are ignored by the tape reader of any numerically controlled machine. However, the punching of a row of delete holes across the tape makes the tape susceptible to tearing.

Any more intimate acquaintance with the Flexowriter must be gained from a practical demonstration and experience.

Later in this book the programming of an elliptical O-ring groove is discussed. Although the periphery of this groove measures only about six inches long, 50 feet of tape or about 6000 characters are required to program the groove. The typing of 6000 characters without a single error is not a small achievement. The person commencing the practice of numerical control must be aware of sources of error and especially his own error patterns. Anyone with experience on office typewriters will be handicapped by the different practice on a Flexowriter. Usually in standard typing the lowercase L is used for the number 1. On the Flexowriter lowercase L is a letter, and a different key is used for 1. It is not enough to know this; inadvertently one may strike L for 1 during tape punching. Another source of error arises from eye movements as the programmer alternately reads the program manuscript and then turns his eyes to the keyboard. More elusive is the suicidal twist in many humans; the author believes that many mistakes are made because of some inner black urge toward minor vandalism. This may be psychological speculation; nevertheless, some keypunching errors seem to arise from unknown reasons. The commission of a keyboard error also may lead to anger, and emotion can be the cause of further mistakes. Since mistakes will be made, it is best to be philosophical about them. Finally, to mention the obvious, never, never type an n/c tape under pressure or the influence of a deadline. A punched tape is either 100 percent correct or 100 percent wrong—if there is any error, it cannot go into the tape reader of the n/c machine.

2.5. THE TAPE CODE FOR PUNCHED TAPE

A variety of tape codes are used in various industries, and it is desirable to employ a single coding system that can serve as an international standard for all information processing. This standard is a new one for the numerical control industry, the ASCII (American Standard Code for Information Exchange), which is the same as the ISO or International Standards Organization code. Until now the character code for numerical control punched tape has been the Electronics Institute of America (EIA) Standard RS244. Both codes are shown in Fig. 16. Some manufacturers supply n/c machines capable of reading either code.

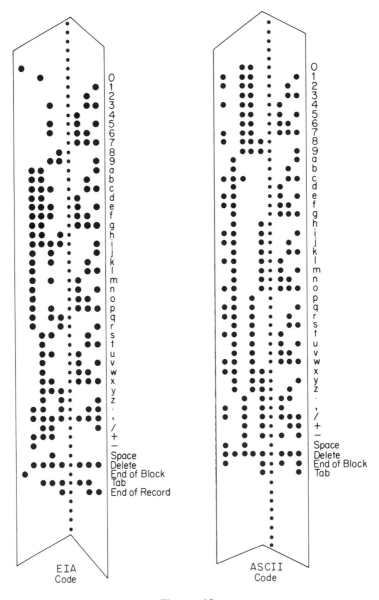

Figure 16

The Tape Code For Punched Tape

Since there are perhaps 10,000 n/c machines using the EIA code, this code will persist for some time. Some machines will be adapted to take the ASCII code, but most probably will not. The EIA code is still being supplied for n/c machines.

Both the EIA and ASCII tape code use the binary coded decimal number system, a variant of the binary number system. The binary number system uses only the digits 0 and 1, a system admirably suited to hardware and software both, since 1 means ON for hardware or "hole" for software such as cards and tape, and 0 means OFF for hardware or "no hole" for software. The following is a quick summary of the binary number system:

Decimal number	Binary number
0	0
1	1
2	10
3	11
4	100
5	101
6	110
7	111
8	1000
9	1001
10	1010
16	10000
32	100000

In the binary-coded decimal system, each digit is coded in binary rather than the whole number. Thus in binary-coded decimal (BCD), the decimal number 32 is coded binary 3 binary 2, or 00110010, rather than binary 32, which is 0100000.

The eight tracks of the EIA tape are organized as shown in Fig. 17. The binary levels 1, 2, and 4 are on one side of the drive sprocket holes.

The coding of the digits 0 to 9 is shown also in Fig. 17. It will be noticed that the numbers 3, 5, 6, and 9 have an additional punch in the fifth or CH track. When these numbers are punched (without the fifth channel) there is an even number of holes, all other numbers having an odd number of holes. Tape punching equipment is remarkably reliable; nevertheless, there is the possibility of a hole's being unpunched or blocked by debris. In the EIA standard there must be an odd number of holes for every character. If one of the punches should fail, then an even number of holes would result, and the EIA tape reader is designed to stop reading the tape if an even number of holes should be read for any char-

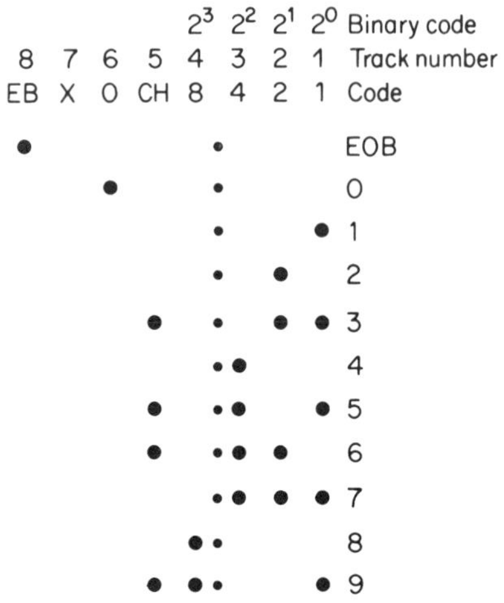

```
                  2³ 2² 2¹ 2⁰  Binary code
        8  7  6  5  4  3  2  1  Track number
        EB X  0  CH 8  4  2  1  Code
        •           •           EOB
              •        •           0
                          •   • 1
                             •   2
                       •  •  •  • 3
                          • •      4
                    •     • •   • 5
                       •  • • •   6
                          • • • • 7
                          • •     8
                    •  •  •     • 9
```

Figure 17

acter. This check system is referred to as *odd parity check*. The ASCII/ISO code uses even parity, with the eighth track the parity track. Tape systems other than those for numerical control may use even, odd, or no parity check.

Track 6 in the EIA system is the zero track, and zero is represented by a hole in this track. Track 7 is the X track, and the eighth is the EOB or End of Block track. The EOB code is punched at the end of every line of punched information. To explain, suppose a block of n/c information comprises an X, a Y, and a Z dimension. This block of information would be typed and punched

   X00100   Y06625   Z04125   (EOB)

where EOB means here the End of Block or Carriage Return. The Carriage Return key on the Flexowriter returns the carriage of the typewriter to begin a new line. This Carriage Return code is a hole in the eighth track or End of Block of Information. The numerically controlled machine takes no action on the information read from the tape until it has read the EOB hole.

The EIA and ASCII coding of numbers and letters is given in Fig. 16. In the EIA code currently in use, the letters A to I form a first group, with the X and 0 tracks punched: A as 1, B as 2, C as 3, to I as 9. The odd parity check must also apply. Letters J to R form the second group of letters, using an X punch and number coding as with the group A to I. The letters S to Z are coded with a hole in the 0 track.

The special symbols plus, minus, delete, end of block (carriage return), end of record, and TAB are also shown. The TAB key is used as in ordinary typing, to leave a space between columns of data, that is, to "tabulate." The end of record or EOR code is sometimes used at the beginning of a punched tape. At the end of the n/c operation, the tape may be rewound in the tape reader ready for the next part to be produced. This code character stops the rewinding operation.

The questions on coding at the end of the chapter should be worked through in order to acquire some familiarity with the tape code. It is not necessary to memorize the code, but a casual familiarity is necessary for the many situations where the operator or programmer must scan a tape to find a certain block of information, word, digit, or mistake.

In order to understand the language of information processing, including n/c information, the following definitions should be understood.

A *bit* is the basic unit of information, either a hole or no-hole. "Bit" is an abbreviation of "binary digit."

A *word* is a unit of information, such as a dimension (X01000 or Z10015), or EOB, delete, etc.

A *block* is a complete group of information words, such as

    H003   G81   X01000   Y06675   M00

This particular block contains five words. Every block of information must, without exception, and for every n/c machine, be terminated with an EOB punch.

2.6. TAPE FORMATS

The compiling of n/c data into suitable blocks of information for action by the n/c machine control unit follows either of two standard formats: *word address* and *tab sequential*. A third format, *fixed block,* is used with the Moog Hydrapoint mills and the occasional special n/c application. Though unusual, the fixed block format has some special advantages, and it and the ingenious Moog machines are discussed in Chapter 5.

2.7. WORD ADDRESS FORMAT

Most n/c machines use the word address format. For an example of this format, consider a complete block of information for the well-known two-axis Cintimatic vertical-spindle mill of Fig. 18. A typical block might read

    H015   G81   X07875   Y04375   M06   (EOB)

There are five words in this block, each preceded by its letter address. The H word is identified or addressed by the letter H and is switched into the H register or storage. Similarly, the preparatory or G word is switched into the G storage. The X, Y, and M words are likewise identified and stored in the proper memories. The block is completed with the EOB

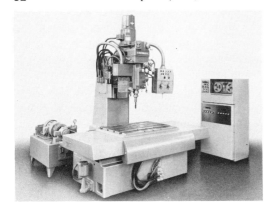

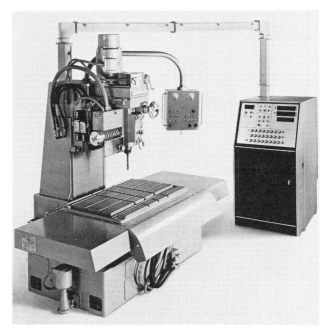

Fig. 18. A display of the family of n/c mills produced by Cincinnati Milacron. (a) The older model of vertical-spindle two-axis Cintimatic mill, used in many colleges for training in numerical control. This machine is used to demonstrate the principles of positioning control in Chapter 3. (b) The newer version of the Cintimatic vertical-spindle mill. (c) The Cintimatic horizontal-spindle mill. The two-axis version is programmed like the vertical-spindle mill. This machine is discussed in Chapter 6. (d) The Cintimatic turret drill, discussed in Chapter 6.

Word Address Format

(c)

(d)

punch. This is a signal to the machine control unit to release the tape information read into storage and to operate with it.

Most n/c machines use a five-digit dimension word for X, Y, and Z. Some use six digits; the Gorton Tapemaster uses four. With five digits, three are decimal places, or with six, four are decimal places. Hence, the above dimensions are $X = 7.875$, $Y = 4.375$. The decimal point is not punched into the tape.

Some machines allow the omission of trailing zeros, and a few allow the omission of leading zeros, but this must never be assumed. If such omission is possible, this will be stated in the programming manual of the machine. Thus, if the dimension is 4.36, omission of leading zeros would permit the punching of 4360, or omission of trailing zeros would permit 0436. When in doubt, all zeros can be included for any machine using any format. Omission of zeros, however, reduces the tape length and probably the number of errors.

It must be clearly understood that omission of leading or trailing zeros is allowable *for dimension words* only. The sample block of information for the Cintimatic includes an M06 (an instruction to stop the machine for a tool change). If omission of leading zeros were allowed (actually not possible for the Cintimatic), this word, not being a dimension word, must be typed M06. If M06 is required, but M6 is punched, the tape reader of any machine will stop. Similarly, H015 cannot be punched H15.

To align and space out the words on the typed manuscript that is produced with the punched tape, it is allowable for almost all n/c machines to punch the TAB key between words. The TAB key moves the carriage of the typewriter over to a preset stop for tabulating columns of data. The TAB code in the tape is ignored by the machine control, just as the DELETE code is ignored.

A word read from the tape into the memory of the control unit will remain in the memory until replaced by another word, or until the end of the program. This principle holds for all n/c machines. There are certain rather obvious exceptions to this principle; for example, M00 is an instruction to the machine to stop, and this cannot be retained in memory, for if it were, the machine would stop at the end of each block of information. Retention of words in memory results in shorter tapes. For example, suppose a row of holes must be drilled on the line Y06000; thus,

X01000Y06000
X02500Y06000
X04000Y06000
X05500Y06000
X07500Y08000

The Y word changes for the last hole.

Fixed Block Format

The word Y06000 remains unchanged and therefore need not be repunched, either for word address or for tab sequential. It may be repunched if desired. This program can be punched

>X01000Y06000
>X02500
>X04000
>X05500
>X07500Y08000

2.8. TAB SEQUENTIAL FORMAT

To illustrate tab sequential format, suppose we recall the previous example of Cintimatic word address format:

>H015 G81 X07875 Y0475 M06 (EOB)

If the MCU is designed to receive a tab sequential format, each word would not be addressed, but would be preceded by the TAB punch. Here we use the symbol T to designate TAB. The tape then would read

>T015 T81 T07875 T04375 T06 (EOB)

The MCU uses a stepping switch to put these words into storage. It reads the first T word and switches it to the first storage position. On reading the second T, the switch advances one stepping position and puts the second word into the second storage position. The stepping switch similarly switches the remaining words into storage.

Again, words that are already in storage do not need to be repeated. Suppose the program reads

>T015 T81 T07875 T04375 T06 (EOB)
>T016 T81 T07875 T05625 T09 (EOB)

The 81 and 07875 words are not erased from storage by new words, so that the program may be punched

>T015 T81 T07875 T04375 T06
>T016 T   T   T05625 T09

Each line of information must, of course, have the required number of TAB's punched. If the second block were punched

>T016 T05625 T09

then the machine control unit would attempt to put 05625 into the G storage position and 09 in the X storage. But since 05625 would not be a recognizable G word, the tape reader would stop reading.

2.9. FIXED BLOCK FORMAT

The fixed block format used with the Moog Hydrapoint (and possibly other machines) requires that a fixed number of characters be included in each block of information. The words are strung in continuous sequence without being addressed or TAB-coded. This format may be illustrated by coding the previous example, which was

        H015   G81   X07875   Y04375   M06   (EOB)
        H016   Y05625   M09   (EOB)

In a fixed block format this must be punched

        01581078750437506   (EOB)
        01681078750562509   (EOB)

If this happens to be the fixed block format acceptable to the machine (it is not acceptable to the Hydrapoint, which among other things uses a one-digit G function), then *every* block must have 18 digits, not including the EOB punch. The MCU reads the first three digits into the H storage, the next two into G storage, and so on. In this arrangement the number of digits in the block cannot be varied.

The disadvantage of this method is the longer tape that results. For positioning work, this may not be significant, but contouring operations would require tapes of unusual length.

2.10. MISCELLANEOUS FUNCTIONS

Whatever the tape format used, most tape codes use certain standard miscellaneous functions or M functions. These must be familiar to programmers.

- M00   Program stop. This is programmed if the operation is to stop for an inspection, the insertion of a component part, or any other reason. The operator presses the CYCLE START button when he is ready to continue the operation.
- M02   End of program. This also causes the tape to be rewound in some machines. Coolant is turned off. Memory is cleared of all words.
- M03   Spindle clockwise rotation.
- M04   Spindle counterclockwise rotation.
- M05   Spindle off.
- M06   Stop for a tool change. Spindle stops rotating.
- M07   Mist coolant on.
- M08   Flood coolant on.
- M09   Coolant off.

Not all machines use all these M codes. Some use only M08 for coolant on, without offering a choice of flood or mist.

Questions

1. Why does the character DELETE have seven holes in EIA and eight in ASCII?
2. Could zero be coded as no holes in any of the tracks in either code?
3. Summarize the system of coding the 26 letters in the two codes.
4. The miscellaneous functions M00, M02, and M06 are all stop commands. What is the difference between them?
5. Consult the EIA character code and read the following EIA tapes.

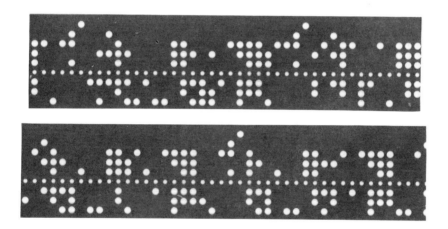

Figure 2-1

6. Suppose the holes punched into the EIA tape (including the parity holes) represented binary numbers. What binary number would be given by the hole pattern for
   a. EOB?
   b. TAB?
   c. DELETE?
   d. 5?
   e. 9?
7. You have punched an EOB in the wrong place in the tape. Can you use the DELETE punch to cancel it? If you are not sure what happens here, make up a short test tape with a deleted EOB and try it.

8. When feeding punched tape through a tape reader, cumulative error allowed is ±0.009″ in 6 inches of tape. Hole spacing is 0.100±0.002 inch in both directions. A sprocket drum with 24 sprocket points is to be designed for driving punched tape. Determine drum diameter, diameter tolerance, and angular tolerance between sprocket positions.

PART 2

MANUAL PROGRAMMING METHODS

3

PROGRAMMING IN TWO AXES

3.1. BASIC KNOWLEDGE FOR NUMERICAL CONTROL PROGRAMMING

Many institutes and colleges have selected for their numerically-controlled machine tool either a Cintimatic vertical spindle mill or Slo-Syn controls fitted to standard vertical spindle mills such as the Bridgeport. These two n/c systems will be used in this chapter to illustrate the basic methods of programming. The Cintimatic machine uses word address format and absolute dimensioning; the Slo-Syn uses tab sequential format and incremental dimensioning. Familiarity with these two systems therefore gives the student of numerical control the basic knowledge that equips him to understand the programming of any standard numerically-controlled machine. The student's n/c machine may be neither the Cintimatic or the Slo-Syn. If it is perhaps a Compudyne Contoura, he must of course learn the programming methods specific to that machine. But his objectives surely are broader than the programming of one machine only, since this narrow approach has obvious disadvantages. The one machine at his disposal is simply a sample of the whole range of n/c machines.

The basic abilities required of a practitioner of numerical control programming must include the following:

1. Ability to program and correct a tape in either word address or tab sequential format.
2. Ability to read a tape in either format.
3. Ability to set up the workpiece and to preset tools for a program on at least one machine.
4. Ability to contour lines and arcs on a positioning mill. These contouring methods are the subject of the following chapter.

5. Experience with at least one type of standard positioning mill.

These are the basics only. A competent practitioner must also be experienced in the following:

6. Programming an indexing table.
7. Programming automatic tool changers with coded tools.
8. Programming of three-axis machines such as the Kearney & Trecker Milwaukee-Matics.
9. Programming of contouring machines such as lathes and contouring mills in both linear and circular interpolation.

All these subjects are discussed in later chapters.

Finally, if the practitioner has pretensions to expertise, he must learn some computer techniques. This requires the study of some of the computer languages developed for the purpose, such as ADAPT, AUTOSPOT, and APT. Fortunately none of these computer language systems are difficult to learn.

In this chapter the Cintimatic mill will receive emphasis, while the Slo-Syn will be discussed only in a general way as an example of tab sequential programming. More complete details on the Slo-Syn will be found in Chapter 9. In order to make convenient comparisons with the Cintimatic mill, which uses five digits, Slo-Syn movements in this chapter will also be dimensioned in five digits, instead of the usual four digits for this machine.

3.2. THE VERTICAL-SPINDLE CINTIMATIC

A general view of the Cincinnati Milacron Cintimatic Vertical-spindle Mill is given in Fig. 18(a). The Cintimatic has two-axis capability in X and Y and can drill, tap, or mill parallel to the X and Y axes. Z operations are preset and called up on the tape as required. The tape coding for this machine is typical for a number of similar n/c mills.

Only positioning operations are discussed in this chapter. How to contour in two axes on such a machine is deferred to the next chapter.

Specifications for the Cintimatic are these:

1. Spindle horsepower, 3.
2. Longitudinal table travel (X axis), 40 inches.
3. Cross travel (Y axis), 20 inches.
4. Maximum spindle travel, 8 inches.
5. Quill feed rate, 2–30 ipm.
6. Spindle rapid approach, 150 ipm.
7. Spindle feed rate, 1–40 ipm, infinitely variable.
8. Table rapid traverse rate, 200 ipm.

9. Milling feeds, 2–30 ipm.
10. Positioning accuracy, ±0.001 in. (repeatability is even better).
11. Spindle speeds, 85–3800 rpm, infinitely variable.
12. Tape reading speed, 60 characters per second.

The origin of coordinates or zero point is assumed to be at the front left-hand corner of the work table, as shown in Fig. 19. (This is not

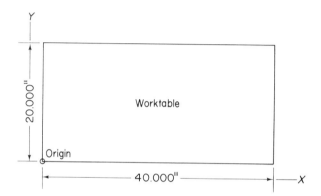

Figure 19

true of all Cintimatics.) All measurements are in the first quadrant; therefore, minus dimensions are not programmed.

Panels of manual controls and indicator lights may be seen on the machine control unit and on the control panel to the right of the machine spindle. These control switches and condition indicators are used for both manual and tape-controlled operation. It is not the intent here adequately to discuss the manual operation of the Cintimatic, even though manual operations are required in setting up a new job and in checking out a new tape.

We turn first to the control panel of the machine control unit; this has master START and STOP pushbuttons to turn on and off the motors, hydraulics, and electronics. Below these are two more pushbuttons, CYCLE START and CYCLE STOP. The tape reader can be stopped at any information block by pressing the CYCLE STOP button. Tape reading continues upon pushing the CYCLE START button. A CYCLE SELECTOR switch is available to provide the following operations under manual control: drill, bore, tap, mill, etc. These are the preparatory or G functions that can be called up by the tape, and will be discussed presently. Another switch gives either manual or tape control of the operation. A search button also is provided. On being pressed, this button will back up the tape to the previous sequence number or block of information. A parity light near the bottom of this panel is useful in checking out a new tape. If this light goes on as the tape is fed through the tape reader, then either there is a parity failure, or more likely some unusual character has been punched into the tape. Finally, zero shift switches can move the origin of coordinates if this is needed.

Only the depth cam elements of the instrumentation on the control panel at the left of the machine tool head will be mentioned here. A bank of disks for nine cams may be seen beside the spindle. These cams are used to set the Z depth for the cutting tools. These cam depth settings can be called up on the tape by designating the cam number, 51 to 59, or a specific cam may be manually selected by turning the cam selector switch on the panel to the required number. When the tool in use is advanced to its depth setting, a cam light goes on.

3.3. SETTING UP THE WORK PIECE

In the section showing several numerical control part prints there is a set of five prints for the manufacture of a small gear pump. The prints show only the machining operations required on the gear pump housing parts; the pump must also have a pressure seal, bushings, locating dowels, and a pair of gears mounted on shafts.

The manufacture of this small gear pump by the methods of numerical control will be extensively discussed throughout the pertinent chapters of this book. All the machining operations required to make the pump housing, including the contoured O-ring groove, are possible on a two-axis positioning mill such as the Cintimatic. Only the drilling and reaming operations will be dealt with in this chapter.

Suppose that we limit our attention to the drilling and reaming of the two dowel holes in gear housing no. 1, drawing no. 1.

The gear housing is made of mild steel, or, better still, a high-strength, low-alloy structural plate such as T-1 or Cor-Ten, 1 in. thick. Dimensions are $2\frac{3}{4} \times 3\frac{5}{8}$ in.

Positioning accuracy of the Cintimatic and most other similar n/c mills is 0.001 in. This accuracy is obtainable only if the work piece is accurately affixed to the work table of the machine, and this is not possible unless the four edges of the gear housing blank are shaped or planed smooth and at right angles to each other.

The blank to be machined may be held by milling clamps or by a heavy and rigid milling vise. A light vise with a moving jaw that rotates slightly under pressure cannot be used for affixing pieces to be machined by numerically controlled methods.

A suitable location for the milling vise holding the work piece is somewhere toward the middle of the table surface. Suppose it is decided to locate the lower left-hand corner of the work piece at $X = 10.000$, $Y = 03.000$. This point is termed the *set-up point*. It rarely corresponds to the zero point or origin of coordinates. The work piece can be set up to give approximately this location, but the first problem to be met in n/c machining is to locate at *exactly* the set-up coordinates desired.

The methods for accurately locating the work piece really must be demonstrated, but perhaps can be sufficiently explained. Some work tables are drilled to receive accurate locating dowels at known XY locations. If the work piece or fixture is located and clamped against these dowels, then its location is accurately known. Sometimes an accurate series of locating dowel holes is drilled into a fixture called a subplate, which is laid on top of the work table and is itself accurately located on the work table. Still another method is to locate a hole at a desired set-up point, and by using the familiar wigglers or center-finders for this purpose, carefully to juggle the part into exact location.

The best method is perhaps an optical method. Suppose a corner of the work piece must be located at X10000 Y03000. The operator affixes the part at that location as closely as he can. He then moves the spindle of the machine to this location by dialing in these dimensions on the manual X and Y switches on the machine control unit. The spindle can be located accurately at any location, and the method followed here locates the work piece from the located spindle. An optical device is inserted into the spindle. The operator looks into its eyepiece to sight on the point of the part to be located, and if this point appears exactly at the intersection of the cross hairs of the optical system, it is accurately located at the correct X and Y location. Almost always it is not.

The part may then be tapped or otherwise adjusted into position. Or the *zero offset capability* of the n/c machine may be used. Suppose that the part corner used for locating is displaced from the desired position by a few thousandths in both X and Y. The zero offset dials may be used to move the origin a few thousandths until the part is accurately in position. That is, the coordinate system is shifted to receive the work piece at the desired coordinates.

The part is now accurately located at a known X and Y position, X10000 Y03000, and is ready for machining.

3.4. THE INFORMATION BLOCK

A full block of information for the Cintimatic contains five words:

Word	Address	Digits
Sequence number	H	3
Preparatory function	G	2
X position	X	5
Y position	Y	5
Miscellaneous function	M	2

The program manuscript form is shown in Fig. 20.

```
                    PROGRAM SHEET
PART_____                    DATE_____
                                        PROGRAMMER_____
PART NO._____ DWG NO._____     SHEET___of___SHEETS
```

SEQ	G	X	Y	M	REMARKS	

Figure 20

Note that any TAB codes in the tape, like DELETE, will be ignored by the machine.

3.5. SEQUENCE NUMBER

The H word is simply a sequence number. That is, the first block of information is designated H001, the next H002, the third H003, etc. However, if there is any possibility of additional blocks being included in the tape due to later modifications, a series of numbers may be left out. Or if the program is entirely exploratory, before settlement of final design, sequence numbers might run 005, 010, 015, etc.

The sequence number is displayed on a sequence number readout on the machine control unit. By this means the operator knows at all times where in the program the machine is working. But why should the operator have to know this if the process is automatic?

Actually, there is no truly automatic process anywhere; most systems are as automatic as circumstances and economics allow. Suffice it to say that there are many situations that call for the intervention and adjustment of the machine by the operator, and he cannot control the operation without the assistance of the sequence number readout to guide him. This readout is especially useful when one is checking the tape by running it through the tape reader in a diagnostic run performed without operating the machine.

It is usual to code the sequence number with an H for the more important blocks of information, and with an N for a partial block of information with some words omitted. The H blocks must be complete with all words. The first block in the program and all blocks that include a tool change should be full blocks, coded H. The MCU has a tape rewind control button, which, if pushed, winds the tape back to the last H block. The operator can thus search back to any portion of the tape if this should be necessary. All blocks may be coded H, but then to search back, the rewind will stop at every block, which makes a tape search tedious. All blocks except the first could be coded N, but this makes a tape search impossible. Experience quickly demonstrates that a tape without H codes is not a tape that can be readily "debugged."

3.6. MISCELLANEOUS FUNCTIONS

The X and Y words are five-digit numbers, three decimal places. Since all locations are in the first quadrant, in the case of the Cintimatic, no minus numbers are used and no minus signs are needed before dimensions.

The few M functions used with the Cintimatic are standard for other machines also, except for M51 to M59.

1. M00. *Programmed stop.* This code stops the tape reader and machine at the end of the block of information in which this function is included. The spindle retracts. The program may be continued by pressing the CYCLE START button.
2. M06. *Tool change.* This is a stop also, with the addition that the tool change light comes on. Spindle rotation is stopped. As usual, the stop is made after the other commands in the information block are executed.
3. M02. *End of program.* Spindle and coolant stop and tape is rewound to the start of the tape program if the machine is equipped with tape rewind capacity. This may require that the tape program begin with EOB. All information in storage is erased. As usual, this command is executed after the other commands in the information block are executed.

4. M50. *No cam.* For explanation, consult the following commands.
5. M51 to M59. Depth cams for controlling tool Z depths are preset before tape reading and machining begin. These cams are numbered 51 to 59 (or some other series of numbers). Punching of one of these numbers calls up the depth setting previously set up on this cam. This procedure will be further explained below. The bank of cams may be seen on the left-hand side of the spindle.

3.7. A LOOK AT THE PART PROGRAM

So far we have discussed a welter of small programming details, and more must yet be discussed. Fortunately, these matters are relatively easy to remember and to use, but it might be wise at this point to give them significance by examining the part program for drilling the two dowel holes in gear housing No. 1.

The holes to be drilled may have to be started with a spot drill or other type of starter drill. We shall assume that this is not necessary, since spiral-point or other centering drills will be used. Both holes must be drilled and reamed to 0.3750, so a $\frac{23}{64}$ drill is selected, to be followed by the reamer. The drill is inserted into the spindle before the program is commenced. The depth of the reamed holes is determined by the length of the dowels. But in setting the depth cam for the $\frac{23}{64}$ drill, it must be recalled that the tapered point of a drill is 0.3 times as long as the diameter, in this case $\frac{1}{8}$ in. long. This influences the depth setting.

At sequence number 001, the program calls for a G81 drill cycle, to be discussed presently, with hole location at X and Y coordinates as given. See Fig. 21. The depth of drilling is set on cam 51, which is called out as M51. The block of information is punched

H001   G81   X11375   Y06280   M51   (EOB)

At sequence 002 a hole is drilled at new coordinates, with depth control by the same cam 51. The callout of cam 51 is still retained in the memory of the machine control unit, and all vertical movements of the spindle will go to the depth setting of cam 51, until another cam number, with a different depth setting, is called for.

As soon as the hole of sequence 002 is drilled and the tool retracted, the tape reader stops and the tool change light comes on. The operator removes the drill from the spindle and substitutes the reamer. He then presses the CYCLE START button, and the tape reader reads the next sequence 003. Notice that a full information block is programmed as

PART__________ SHEET NO.___ of ___
REVISION__________ PROGRAMMER__________
PART NO.__________ TYPIST__________
MACHINE__________ DATE__________ RELEASED__________

SEQ	*G*	*X*	*Y*	*M*	REMARKS
H001	81	11375	06280	51	23/64 Drill
H002	81	11375	03470	06	0.375 Reamer
N003	85			52	
N004			06280	02	

Figure 21

an H sequence. A tool change is an important block, and one might want to search back to it.

The spindle is still located at X11375 Y03470, so that this hole is reamed first. The coordinate dimensions are already in storage, and are not repeated on the program sheet. The depth setting for the reamer is different from the drill depth setting, so this new depth is set on cam 52. The G85 preparatory cycle is a bore cycle, to be discussed.

In the last sequence, G85, X11375, and M52 are recalled from the memory and thus not recorded on the program sheet. The program must end with M02.

3.8. MANUAL CONTROL OF Z-AXIS MOVEMENTS

The cams numbered 51 to 59 are actually pairs of cams on which Z-axis depths are preset before the operation begins. Calling up the cam number on the tape puts the preset depths into the program. One cam of the pair sets the rapid-advance-to-work-surface depth, and the other the depth of feed.

The method of setting the R (rapid advance) cam is explained with the help of Fig. 22. The rapid advance must stop just before the

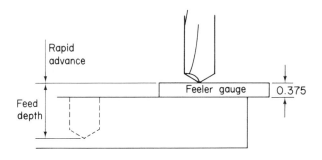

Figure 22

tool comes in contact with the work piece; the accepted distance from the work piece is 0.100 in. But since the rapid advance movement tends to overrun this setting, the R distance is set back an additional 0.275 in. When one is setting up the R position, then, the tool is manually advanced until it touches the 0.375 feeler gauge, as shown in the figure. The cam is adjusted at this position so that the cam light goes on, indicating the required setting.

To set the F (feed) cam to give the proper depth of feed, the tool point is manually positioned down against the top of the work piece. The F cam setting is adjusted so that in this tool position the cam light goes on. The cam setting is then moved the additional distance required to go to actual tool depth in the work piece.

The rapid advance setting applies to a specific length of projection of the tool from the toolholder. If, when one is using the same cam, the tool is replaced by a slightly longer tool, the cam setting will ram the longer tool into the work. Frequently in n/c operations the tool projection is preset on a presetting gauge. Rough and finish end mills are often preset to different projections so that the same cam may be used for both.

Giddings and Lewis provide n/c drills that sense when the tool touches the part. This pressure sensing is used to change the Z motion from rapid advance to feed rate.

3.9. THE PREPARATORY FUNCTIONS

This section explains the sequence of motions for each of the preparatory functions. These G functions are also illustrated in Fig. 23.

G78. MILL CYCLE STOP. The work table moves at rapid traverse to the programmed XY position in the information block. The quill then moves down at rapid advance and slow feed rate to the final depth setting. The operation then stops and the quill-clamp light goes on. The operator clamps (or unclamps) the quill. Then he pushes the CYCLE START button to read the next block of tape.

This G function is used when a milling cutter must be fed into the work to depth before commencing to mill. The same function may be used at the end of the milling cycle so that the operator can unclamp the quill and retract it. If used at the end of a milling operation, the cutter is, of course, already at final depth, so that there is no Z motion of the tool; in this case G78 functions more or less as a stop code with clamp light on. It is best to clamp the quill even though a uniform depth will be maintained by the unclamped quill.

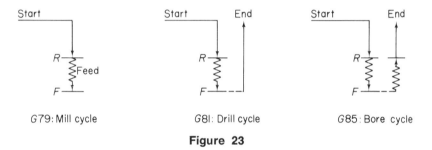

Figure 23

Two-lipped end mills, also called slot drills, must be used for feeding into the work in the Z direction.

G79. MILL CYCLE. The work table moves to the programmed XY position in the block at the cutting feed rate (under G78 it moves at rapid traverse). The spindle then goes through Rapid and Feed depths and stays down. The next tape block is immediately read. This command is used after G78 to commence milling. There is no Z movement from this command if the tool is already at the full depth due to a G78 in the previous block.

G80. CANCEL CYCLE. The machine moves at rapid traverse to the XY position given in the information block, but there is no Z movement. The next block is immediately read. This command is used for movements with the quill retracted (as when moving around a projecting clamp or fixture), for rapid positioning, or to cancel out a previous cam setting.

G81. DRILL CYCLE. The work table moves at rapid traverse to the XY location programmed in the block. The spindle then moves through rapid traverse and slow feed to the final depth setting. The spindle then retracts rapidly.

G84. TAP CYCLE. The work table moves at rapid traverse to the XY position programmed. The spindle then moves through rapid traverse and slow feed to the final depth setting on the cams. The spindle then reverses and then rises at slow feed rate to the feed engagement point. The spindle then returns at rapid traverse and reverses again.

G85. BORE CYCLE. This sequence is identical to G84, except that the spindle does not reverse. Thus the boring (or reaming) tool is removed from the hole at feed rate while still rotating, to avoid marring of the finish of the hole. If finish is not important, boring may be done with a G81 command.

Note that all G, X, Y, and M cam numbers remain in effect until a new command is given. However, M00, M02, and M06 are not retained in memory. All XY moves are made before any Z motion occurs.

If a dwell is required when a boring or counterboring tool has reached final depth, the dwell is programmed by a series of TAB's or DELETE's. Since tape reading speed for the Cintimatic is 60 characters per second, each such character will produce a dwell of $\frac{1}{60}$ second. The dwell may be programmed in a number of ways, one being the following:

>G80   X10000   Y10000
>G79
>(TAB TAB TAB TAB TAB etc.)
>G81

In this example, the tool is positioned rapidly at the boring location with a G80 command and goes to depth with a G79 command, or a G78. The tool is withdrawn by the G81 command. All three G functions are called at the same location, X10000 Y10000.

3.10. PROGRAMMING CONSIDERATIONS

In these descriptions of numerically-controlled operations, the tool or spindle is described as moving to the XY point designated in the information block. Now of course the worktable, not the spindle, moves, whether the system is Cintimatic, Slo-Syn, or Contoura (though the spindle moves in the case of some machines, such as the Milwaukee-Matics). Nevertheless, in numerical control programming, the assumption is always made that the *spindle* moves, not the worktable. The reason for this assumption is largely that computer programming is based on this assumption.

There are some weaknesses to be examined in the program of Fig. 21.

A tool change is made while the cutter is positioned above the workpiece. As a protection to both tool and workpiece, it would be preferable, and may sometimes be necessary, to move the spindle to some location away from the workpiece before changing the tool.

Again, at the end of the program, the spindle is located over the workpiece. If the workpiece is removed when the spindle or tool is above it, there could be an accidental collision between the two. The spindle should be removed from the area of the workpiece at the end of the program. This could of course be done by means of the manual controls.

3.11. TAB SEQUENTIAL PROGRAMMING

The same two dowel holes will be programmed for the Slo-Syn controls, which use tab sequential format and incremental positioning.

An incremental movement from the previous position will be either a plus movement, or a minus movement in the opposite direction to a plus movement. These sign conventions will be more thoroughly discussed in later chapters, but can be summarized thus:

1. A spindle movement toward the right of the operator is $+X$; a spindle movement toward the left of the operator is $-X$.
2. A spindle movement toward the operator (forward) is $-Y$; a spindle movement away from the operator (to the back of the machine) is $+Y$.

In writing a Slo-Syn program for the two dowel holes, suppose we adopt a parking position for the tool at X00000 Y00000. This is not perhaps the best choice of position. The parking position is the initial position of the tool at the commencement of the program, the final position, and the tool change position.

The Slo-Syn information block includes four words thus (T represents the TAB key):

(seq. no.) T (X increment) T (Y increment) T (M word) (EOB)

For the Slo-Syn or any other tab sequential n/c machine, there must *not* be a TAB character before the sequence number, because the sequence number is not a machine instruction, but merely a display.

The controls are set up so that a hole is drilled at each location programmed, except when M06 or M02 is programmed. In effect, a G81 Drill Cycle is preset on the machine and is not needed in the program. The Slo-Syn does not employ the "canned" G cycles used with the Cintimatic. Here is the program:

```
                        EOB
0 RWS                   EOB  (tape Rewind Stop)
1 T  11375  T  06280    EOB  first hole
2 T          T—02810    EOB  second hole
3 T—11375 T—03470 T06   EOB  parking position
```

A tool change is made at the parking position. Note in sequence 1 and 2, that the third TAB can be omitted (or it can be included) because no further information is switched into the machine. Words already in storage need not be repeated.

```
4 T  11375  T  06280        EOB
5 T          T—02810        EOB
6 T—11375 T—03470 T02       EOB
```

An M02 ends the program. The Slo-Syn controls are actually positioned with four digits rather than the five digits used here for the sake of parallel programming with the Cintimatic.

Questions

1. Counterboring with a dwell of two revolutions at 600 rpm spindle speed must be done at two locations, X12650 Y06625 and X16125 Y09750. Write the program for the Cintimatic vertical-spindle mill.
2. A slot is to be milled parallel to the Y axis at X06600 from Y03500 to Y07500 (these positions are initial and terminal positions for the milling cutter).
   a. Since only a light cut is to be taken in aluminum alloy, the quill will not be clamped. Write the program.
   b. Write the program for a Quill-Clamp at start and Quill-Unclamp at finish of the cut.
3. Program the drilling of the four bolt holes in gear housing no. 1 of the gear pump, part print no. 1. The lower left-hand corner of the part is affixed at X10000 Y03000. The Y axis is positive up.
4. Program the drilling of all the holes in the gear spacer of the gear pump, part print no. 1. Include the four holes that shape the large opening in the middle of the spacer. Affixing of the part is as given in Question 3. Assume no interference from clamps.
5. Program the drilling of all holes in the aluminum tooling plate base pad, part print no. 2. The lower left-hand corner of the part is affixed at X10000 Y03000, and Y is positive upward.
6. Program the milling of the small pocket of the base pad, part print

no. 2. The mill is a 1.000-in. slot drill, and since the pocket is 2 in. wide, assume that it can be cleaned out in two passes without overlap of passes. There is to be no finish cut. Affixing of the part is given in Question 5.

Do not start the plunge cut at the pocket periphery, as this can spoil the surface finish. Make the plunge cut somewhere inside the pocket and mill out to the periphery to commence the pocketing. If you do not understand why this procedure is used, try it the wrong way to find out why.

7. Complete the machining operations on the face of the fluidic Schmitt trigger manifold, part print no. 3. The lower left-hand corner of the part is affixed at X10000 Y10000, with Y positive upward.
8. In an n/c machining operation, a milling vise is clamped to the work table to hold the work pieces. What procedure would you use to align the vise jaw so that it is parallel to the X axis of the machine bed?
9. In large organizations, typists often are borrowed from the typing pool. Such an arrangement is commonly disastrous to a technical department of an organization.

An excellent girl operates your Flexowriter, but today she is away sick. You are offered an office typist from the typing pool who has never seen a Flexowriter and worse still, knows that an occasional typing mistake is acceptable. Discuss the significance of the following of her errors, and where applicable, how the n/c machine will react and how the error may be corrected:
 a. Lower case L for 1.
 b. O for zero.
 c. Omission of one TAB in a tab sequential format.
 d. Omission of one TAB in a word address format.
 e. M6 for M06 (Cintimatic mill).
 f. G78 for G79.
 g. G79 for G78.
 h. Omission of a whole block of information in a drilling program.
 i. Interchange of two blocks of information in a sequence of hole drilling on a flat plate; all holes the same diameter.
 j. Y word typed before X word in a word address format.
 k. Y word typed before X word in a tab sequential format.
 l. Omission of a sequence number from an information block.
 m. Omission of the tape leader.
 n. Omission of the cam M word for a series of holes to be drilled.

4

CONTOURING OPERATIONS WITH A POSITIONING MILL

Positioning machines with milling capacity are almost always limited to milling operations parallel to X and Y axes only. Such machines can be, and are, used for contouring work, and the techniques for this sort of operation are the subject of this chapter.

4.1. POCKET MILLING WITH A POSITIONING MILL

Two pocket milling operations are required for the base pad shown in Part Print No. 2. Note that the radius of all corners of the pockets is 0.5 in. If both rough and finish cuts are to be used in producing the larger pocket, the roughing cutter could be a larger diameter than 1 in. and the finishing cutter 1 in. Some types of end mills have a chamfer on the cutting end, and if so then finishing cuts must be overlapped if the bottom of the cut is to be flat and not scalloped. Since there is no significant backlash in n/c machines, climb milling is preferred, except for cast irons.

The following procedures program the operation on a Cintimatic. Variations in the procedure are, of course, possible.

To enter the pocket, program a G78 and the XY coordinates where the tool will enter the pocket—at a corner perhaps, if surface finish does not matter. The tool will go rapidly to these coordinates, drill down, stop, and turn on the quill clamp light. The operator clamps the quill and then restarts the tape reading (see the explanation of G78 in the previous chapter). The Cintimatic will cut to a remarkably uniform depth, especially in aluminum, without clamping the quill. However, clamping is to be recommended except for shallow cuts, say, below $\frac{3}{16}$ in.

Next, program G79 and the XY coordinates to which the tool must mill, that is, the termination of the milling cut—the previous G78 was at the start of the cut. G78 calls for milling at feed rate. Continue milling out the pocket over the path programmed.

When the pocket is milled, the quill must be returned, but it must be unclamped first. Program a G78 to allow the operator to unclamp. There may possibly be no other new data in the block except G78. Then in the next block program G81 to retract the quill. This last block may have no new data other than G81 and perhaps M02.

Instead of programming G78 and G79, an alternate method is to program G80, then G79, the same XY coordinates being used for the G79 command as for G80. The G79 command simply lowers the cutter at the XY destination given.

If pocket milling is terminated at a corner and the cutter is then retracted, the cutter will slightly enlarge the corner because of relief of cutting pressure. It is preferable to move the cutter into the pocket away from the periphery of the pocket and then retract the tool. Also, the spindle should be moved to the rear of the work piece at the end of the program for convenience in removing the work piece from the table.

The depth of this pocket can be handled in one cut in aluminum. If the pocket depth must be reached in several stages of cutting in layers, then cams of several depths are required. Also, if the pocket is removed in rough and finish operations, because of differences in the length of the two cutters cams of two depths will be required.

The milling of the small pocket of the base pad will be used as an example of pocket milling. Assume the lower left-hand corner of the part to be set up at X10000 Y03000. A two-lipped end mill is used, 1 inch in diameter. This assumes that the end mill is not undersize. The pocket operation begins by drilling into the center of the pocket to depth. The milling path will be down to the bottom side of the pocket, then around the pocket clockwise.

				EOB
H001	G78	X15000	Y05500	M54
N002	G79		Y05000	
N003		X11500		
N004			Y06000	
N005		X18500		
N006			Y05000	
N007		X14995		
N008	G78			

```
N009    G81
N010    G80    X04000    Y18000    M02
```

4.2. ANGLE CUTS

Suppose that a positioning mill such as the Cintimatic or Moog Hydrapoint must mill a slot at an angle of 60 deg to the X axis. The machine can mill only parallel to X or Y; this slot is at an angle to both. To mill such a slot, it must be approximated by small movements, first in Y then in X. See Fig. 24. Suppose that the allowable bilateral tolerance for the

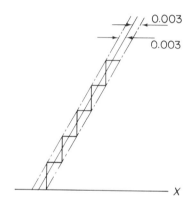

Figure 24

slot is ±0.003, that is, three-thousandths on either side. The tolerance is shown in the figure in the X direction for convenience. To produce the slot, the cutter must mill in the movements shown in the figure. The X movements will be 0.006 in. The Y movements will be 0.104 in., by trigonometry.

If a good finish is required, a tighter tolerance than 0.003 bilateral must be used.

In order to improve the finish by a closer approximation to the desired line, an X and a Y movement should be programmed in each block, not programmed successively. By this means the cutter is caused to move in both axes at the same time.

Positioning machines employ relatively slow tape readers. The Cintimatic, for example, reads the tape at 60 characters per second. Since each block of information provides a cut of only a few thousandths of an inch, the tape reading time will roughly equal the machining time. To reduce tape reading time, the block of information for cutting such a slot should contain only an X and a Y word. Sequence numbers should be omitted.

4.3. THE GENERAL CASE OF THE CIRCLE

Two-dimensional contouring is possible without the assistance of a computer if curves no more complex than the circle are used. Calculations become burdensome if more complex curves than the circle are programmed without computer assistance. Even for the circle, a desk calculator is most advantageous.

For the cutting of circular arcs, three possible cases arise:

1. The cutter must cut on the outside of the circle to produce a convex radius.
2. The cutter must move on the inside of the circle to produce a concave radius or a fillet.
3. The cutter must mill a circular groove, as in the case of the O-ring groove in the prints for the gear pump that follow Part II of this book.

In the third case, the curve is the actual path of the axis of the end mill as it produces the groove. But in the first and second cases, the edge of the cutter just touches the curve, or rather must be within tolerance of the curve, so that the tool axis is *offset*.

Hand programming for the third case only is discussed here. However, the method can be adapted to serve the other two cases by altering the radius of the circle to that required to be traversed by the axis of the cutter. This requires that the radius of the circle be adjusted by the amount of the cutter offset.

We assume here that we must contour, using circular arcs, on a positioning mill such as the Cintimatic. The n/c machine can mill parallel to X and Y axes only. We must approximate a circle by short movements alternately in X and Y, the length of these movements being determined entirely by the allowable tolerance. As in the case of the slot just discussed, a smoother finish is obtained by including an X and Y movement together in each information block.

When a circular groove is being milled, the radius is known and so is the allowable tolerance. The tolerance must be expressed bilaterally, for convenience of computation. Thus, if the radius and tolerance on the drawing are given as $3.200 + 0.10, -0.000$, this must be converted to

$$3.205 \pm 0.005$$

The requirement of a bilateral tolerance applies only to hand programming.

The General Case of The Circle

It is not a requirement of the APT computer programming discussed in later chapters.

In programming the O-ring groove of the gear pump, part print no. 1, the tolerance is a loose ±0.003, if the length of the n/c tape is to be reasonably short. Even with this relatively loose tolerance there are about 600 X and Y calculations to produce the complete groove, and 50 feet of punched tape.

The O-ring groove will be used to illustrate the procedure for hand programming a circle or a circular arc. Tolerance will be 0.003, plus and minus, though a tighter tolerance would be more suitable.

Consult Fig. 25. This shows a circular arc of radius R and bilateral tolerance of n. The end mill commences cutting with the axis of the end mill on the radius at A_0, B_0, where A_0 is presumably an X dimension and B_0 a Y dimension. The origin of coordinates for A and B dimensions is the center of the circle, so that B_0 is actually zero.

The end mill cuts in the Y direction (up the page) for a distance B_1. At this increment in Y it has reached the tolerance limit. The cutter must then move in the X direction until it reaches the inner limit of tolerance. This X increment is $(A_0 - A_1)$. The next movement is a Y increment back to the outer tolerance limit. X and Y alternate as the arc is approximated, with constantly changing increments in both dimensions.

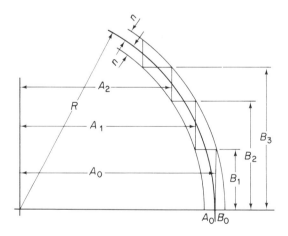

Figure 25

The formulas for calculating these increments are derived from the geometry of the circle.

R = radius
n = tolerance in one direction
  = 0.003 in this example, *not* 0.006

$A_0 = R$

$B_0 = 0$

$A_1 = \sqrt{R^2 - 4Rn}$

$B_1 = \sqrt{2Rn + n^2}$

$A_2 = \sqrt{R^2 - 8Rn}$

$B_2 = \sqrt{6Rn + n^2}$

$A_3 = \sqrt{R^2 - 12Rn}$

$B_3 = \sqrt{10Rn + n^2}$

$A_4 = \sqrt{R^2 - 16Rn}$

$B_4 = \sqrt{14Rn + n^2}$

$A_j = \sqrt{R^2 - 4jRn}$

$B_j = \sqrt{(4j - 2)Rn + n^2}$

Frequently n^2 is infinitesimal compared to the Rn component in the B function and can be ignored. To repeat, the assumed origin of coordinates is the center of the circle.

What are the implications of errors and inaccuracies? Rounding-off errors must be assessed first. Numerical control programming is usually done to thousandths of an inch, which means that "tenths" could be significant when one is rounding off. The fourth decimal place must be watched carefully, but can sometimes be ignored. These calculations, therefore, virtually require the use of a desk calculator or a Fortran program put through a small computer.

Next consider the significance of calculating errors. Suppose that an error is made in the calculation of A_2, and then A_3 is calculated by adding in $4Rn$ more. By this means a single error is promoted into a continuous error.

4.4. PROGRAMMING THE O-RING GROOVE

The formulas for incrementing around a circle in X and Y are simple enough. But the O-ring groove is composed of four circular arcs, and before programming a shape such as this, the programmer must be quite clear at all times as to the relation between A and B increments and the X and Y directions. There is an orientation problem here.

It does not matter where the calculations begin, but it ought to be obvious that you must start at one of the axes of symmetry of the O-ring ellipse. There are then four possible places to begin incrementing, and from any of these points the programmer increments out to the points of tangency where two arcs meet.

Suppose that you were to begin incrementing at the upper end of the vertical axis of the ellipse, that is, the top point of the ellipse, and incremented clockwise (to the right). The vertical direction is Y and the horizontal direction is X. Then A_0 is a Y dimension and B_0 an X dimension. The B's become larger in X and the A's smaller in Y.

You would increment along the top small-radius arc until the point of tangency is reached for the larger-radius arc. The last increment in X and Y must give the point of tangency. Actually, the last two increments are unlikely to give this point exactly. Instead, the actual X and Y positions of the point of tangency are calculated from sine or cosine 30 deg or 60 deg.

The large-radius arc to the right of the ellipse comes next. The programmer is at the point of tangency, not on the intersection of the arc with the ellipse axis. You cannot increment from the point of tangency. On this large-radius arc, A_0 is an X dimension and B_0 is a Y dimension, just the opposite of the case for the small-radius arc.

Since the ellipse is symmetrical, the calculations for any quarter of the ellipse can be adapted to any other quarter.

4.5. PREPARING AND CHECKING THE TAPE

The punched tape for producing the O-ring groove on a typical positioning mill is about 50 feet long, or 6000 characters. A tape of this length offers plenty of scope for making errors. If your experience with the Friden Flexowriter or other tape-punching equipment is limited, you will probably require virtually a full day to punch a correct tape of this length. The commonest errors of inexperience are the following:

1. In glancing from the typewriter back to the written program, the wrong line may be read.
2. The wrong key may be struck.
3. O may be substituted for zero or lower case L for unity.

It is preferable to have someone call the data from the program manuscript while you operate the typewriter. If you make mistakes, do not become emotional; emotions make for further mistakes. If your mistakes become too frequent, stop for a cup of coffee. Never prepare an n/c tape under the pressure of a deadline.

When a punched tape is believed to be correct, it must be run through the tape reader of the machine, without operating the machine tool. Any unusual characters in unusual places in the information block will cause the parity check light to go on.

The final check on the tape program is to make a test run on a piece of aluminum plate or a block of foamed polyurethane.

This O-ring groove is milled with a $\frac{3}{32}$ ball end mill. This is a small and fragile cutter. If you are checking out a tape for such a groove, use a somewhat larger cutter.

In the tape program for incrementing around such a groove, do not punch sequence numbers into the tape, as this adds significantly to the tape reading time. The information block should include only an X and a Y dimension. The standard positioning mills will read the tape and cut this groove with a $\frac{3}{32}$ ball end mill in about $3\frac{1}{2}$ minutes in mild steel.

Questions

1. Program manually the first 10 *A*'s and *B*'s for the top small circle of the O-ring groove ($R = 0.641 \pm 0.003$) for the gear pump, part print no. 1.
2. Write a Fortran program, or have someone else write one, for incrementing around one-quarter of the O-ring groove ellipse, part print no. 1. The tolerance is to be ± 0.003. A tolerance of ± 0.002 will produce a better finish, if you so wish.
3. Fig. 4-1 represents a triangular pocket to be milled. The interior of the pocket has been milled out with a large mill. What remains is to remove the periphery of the pocket with a $\frac{1}{4}$ in. end mill.
   Program this finish cut to a tolerance of ± 0.003. Do not program the whole length of the two sloped straight lines, but instead give the terminal points and the increments in X and Y required to mill these lines within tolerance. Remember—this is a pocket. Offset the cutter inside the pocket.

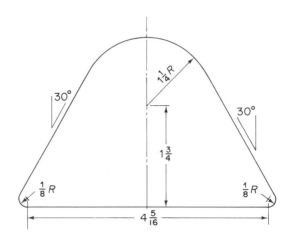

Figure 4-1

4. In contouring, how critical is the matter of rounding off numbers to the nearest thousandth, as between absolute and incremental positioning?
5. Sometimes a flat surface must be milled with a ball end mill. As a result, a scallop is left between passes, as shown in Fig. 4-2. The distance

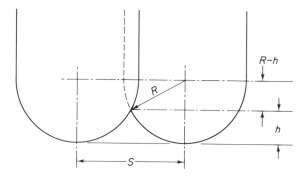

Figure 4-2

S between passes is determined by the allowable scallop height, which is always very small.

Show that $\quad S = 2\sqrt{2Rh}$

5

FIXED BLOCK PROGRAMMING

5.1. THE FIXED BLOCK FORMAT

In the fixed block format the information block must contain a certain fixed number of characters. No letter address or tab codes are used. If such a tape format is to be read by the usual tape reader, then the first three characters would be counted out and switched into the sequence number circuitry, the next two characters perhaps read into the G function circuitry, and so on. Rather than being addressed, the information bits would have to be counted. However, the fixed block format does not require to be read in this manner.

If each block contains the same number of characters, then, of course, all blocks occupy the same length of tape. The Moog Hydra-Point n/c positioning machines use this format, each block containing 20 characters, with a length of 2 in., in the case of the Moog machining center, model 83-1000. If all blocks have the same length, it is most convenient to read the whole block at the same instant, rather than character by character. Each word will occupy a certain location along the tape reader, and sufficient hole-sensing devices can be arranged in tandem to read all characters. This arrangement circumvents the electrical switching equipment required to switch words to the proper storage and reduces storage requirements for the information so read. The format, therefore, is especially adaptable to pneumatic and fluidic methods of tape reading. The Moog machines read fixed block tapes pneumatically.

Since the time required to read the whole block is the time required to read a single character, such an arrangement would seem ideal for contouring tapes, which must be read rapidly to keep the n/c machine supplied with information for the small cuts characteristic of contouring. The format, however, is a handicap for contouring. Since no words or leading or trailing zeros may be omitted, the tape would be at least twice as long as would be required by the standard tab sequential or word address formats. The additional tape length, of course, presents additional scope for tape errors. Fixed block format is confined to positioning work, though contouring is certainly not impossible.

5.2. THE MOOG HYDRA-POINT 83-1000 MC

N/c machines equipped with automatic tool changers are usually referred to as machining centers, since they are capable of executing a wide range of operations on the part in a single setup. This machine is equipped with a tool magazine on the left-hand side, and automatic tool changer (Fig. 26). It is, therefore, referred to as a machining center. It may also be equipped with an automatic indexing table, the Erickson model 400, mounted either vertically or horizontally, to rotate the work piece so that more than one side of the piece can be presented to the tool.

All the Moog n/c machines are modified forms of the famous Bridgeport vertical mill, as other n/c machines have been, and others will be in the future. The machine shows the inventiveness of the Moog brothers, who invented also the Moog synthesizer, famous in musical circles, and the Moog servovalve. No electronics is employed in the machine, the sensing and logic devices being wholly pneumatic and hydraulic. As a result, maintenance is negligible. Tapes may be punched on the machine control unit, a whole block at a time, so that a Flexowriter is not absolutely necessary. It may be noted that error-prone personnel will make fewer errors by punching tapes on the Moog control unit than they will at a typewriter. This method of tape-punching is also advantageous when it is desired to produce the tape at the same time as the first part is being made or designed on the machine. The tape information is manually set up and can be checked for correctness on the readouts before punching. The positioning control method used in this machine is unique, and is another Moog invention. It will not be discussed here. In brief, there is no n/c machine comparable to the Moog units, and their uniqueness gives them some outstanding advantages, with, of course, some compensating disadvantages, such as the nonstandard tape format and very limited contouring capacity.

5.3. THE INFORMATION BLOCK

The information block for the model 83-1000 machining center contains the following words:

Word	Digits	
Sequence number	3	
Preparatory function	1	
X dimension	5	
Y dimension	5	
(Reserved, not used)	1	(punch zero)
Tool code	2	
Miscellaneous function	2	
EOB	1	
	20 characters	

The use of a nonstandard 1-digit preparatory function will be noted.

This block of information controls only two axes, X and Y. The machine is also available with three-axis control. Since the length of the information block must always be 20 digits, the third or Z axis is programmed in the next information block after the X and Y information, the tool rapid advance point (feed point) being used as the X word and the final Z position of the tool as the Y word. The preparatory function 9 indicates to the machine that a Z information block is being read. Thus the machine alternates between two-axis and one-axis control.

In addition to the above 9 function, the following preparatory functions are supplied:

0	Position Only	Work table moves in rapid traverse to programmed XY location. Spindle does not advance.
1	Tap Feed Control	Used for tapping with floating tap holders. Work table moves in rapid traverse to XY position. Spindle advances with rapid and slow feed to final depth, then reverses and feeds out to feed point, then rapidly to zero position.
4	Peck Drill	Table traverses rapidly to programmed XY position. Spindle feeds down rapidly to feed point, then at feed rate until the dwell timer cycles, whereupon the tool retracts rapidly to the feed point. The spindle returns to approximately the previous depth

Fig. 26. Hydra-point machining center (Model 83-1000 MC).

		and continues this "woodpecker" drilling to final depth, then retracting rapidly.
5	Drill	Table moves rapidly to XY destination, then executes the usual canned drilling cycle. Corresponds to G81 function (see Sec. 3.8).
6	Mill	When the spindle is at final Z depth, the work table moves at the preset milling feed rate to the XY position in the information block.
7	Bore	Table moves rapidly to XY destination programmed. Spindle executes a standard boring cycle. See G85 function in Sec. 3.8.
8	Tap	Lead-screw tapping function with spindle reverse at depth.
9	Read Z	No spindle or table motion. Supplies feed point and final depth information. This information holds for following operations on the tape until changed.

The spindle must be retracted when reading a 9 command.

The miscellaneous functions 00, 02, 06, 07, 08, and 09 are the standard functions discussed in Sec. 3.5 and elsewhere. M06, however, is a manual tool change. In addition, other M functions are employed with this n/c machine:

80   No change in miscellaneous function.
81   Partial spindle retract. Limits spindle retraction to feed point only. This command is cancelled by 00, 02, 06, 86, and 82 miscellaneous functions. It must not be used with peck drilling and tapping operations.
82   Partial spindle retract off. Cancels 81 and causes full retraction of spindle.
83   Index Erickson Indexer one angular increment. There may be 4, 6, 8, 12, or 24 increments per revolution.
86   Automatic tool change. The tool function code in the same information block determines the next tool to be used. The tool change is made after movement to the designated XY position in the block.
84   X feed rate change. Signifies that the value in the tool function code designates an X feed rate. See the following chart.
85   Y feed rate change. Signifies that the value in the tool function code designates a Y feed rate. See the following chart.
88   Spindle speed change. Signifies that the value in the tool function code designates a spindle rpm. See the following chart.
89   Z feed rate change. Signifies that the value in the tool function code designates a Z feed rate. See the following chart.

The functions 84, 85, 88, and 89 refer to the following coding:

Tool code	84 (X) or 85 (Y) ipm	89 (Z) ipm	88 rpm
01	1.0	1.0	65
02	1.9	2.1	75
03	2.7	3.6	91
04	4.0	5.3	105
05	5.5	8.1	120
06	7.2	10.3	140
21	—	—	525
22	—	—	625
23	—	—	760

The tool code, therefore, serves as a kind of "variable word address" word. As an example of the use of this tool code table, suppose that an rpm of 760 is desired for spindle speed. Then 23 is used as a tool code number, with miscellaneous function 88. If a milling speed of 4 ipm in Y is required, the tool code number is 04 and the miscellaneous function is 85.

5.4. A PROGRAMMING EXAMPLE

Part print no. 3 is a fluid Schmitt trigger manifold to be drilled for air passages. The 8 holes 10–32 UNF in the four sides of the manifold must be drilled and tapped. For these operations the manifold is mounted with its flat face against the Erickson indexer and can be rotated about a horizontal axis. The indexing increment is assumed to be 45 deg, so that two indexes will be required to turn each side of the manifold face up for drilling.

The drill is preset for length such that a feed point of 1.300 in. in Z results, when side 1 or side 3 is uppermost. That is, the spindle must advance this distance to begin drilling at the feed point, which is 0.100 in. above the surface of side 1 or 3. The setup arrangement is sketched in Fig. 27.

The drilling program for the four sides is given in Fig. 28. The Z final depths would not be finally decided until a trial part was completed and inspected.

The following programming details for the several Moog machines should be noted:

1. The Z depths coded for a 9 function hold for following holes unless superseded.
2. The tool function is coded 00 for functions 00, 02, 07, 08, 09, 80, 81, and 83.

Questions

1. Program all the drilling operations on the gear spacer plate, part print no. 1, using the Moog machining center without Erickson indexer. Locate the lower left-hand corner of the spacer plate at X03000 Y10000 with Y positive up the page. It is sufficient to code fictitious tool code numbers in the absence of the whole tool code table. Use an arbitrary feed point of 1.000 in. for all tools.

A Programming Example

2. Program the face of the fluidic Schmitt trigger manifold. Locate the lower left-hand corner of the manifold at X06000 Y06000 with Y positive up the page. Use an arbitrary feed point of 1.000 in. for all tools.

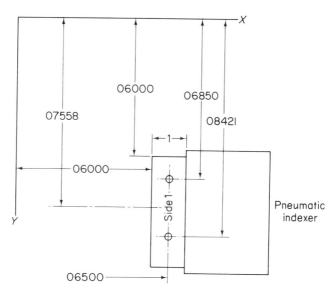

Plan view

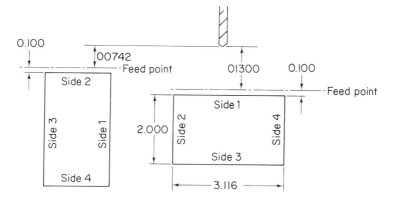

Elevation views

Figure 27

MOOG INC.

Program sheet
HYDRA-POINT

SHEET __ OF __

TAPE NO. _____ ZERO OFF SET X _____ Y _____
PART NO. _____ REV. _____ PREPARED BY _____ DATE _____
PART NAME __Fluid manifold__ TYPED BY _____ DATE _____

DESCRIPT'N	SEQNCE NO.	PREP. FCT.	FEED POINT X-AXIS 00.000	FINAL DEP. OR Y-AXIS 00.000	RES FCT.	TOOL FCT. CODE	MISC. FCT.	DESCRIPT'N	SEQNCE NO.	PREP. FCT.	FEED POINT X-AXIS 00.000	FINAL DEP. OR Y-AXIS 00.000	RES FCT.	TOOL FCT. CODE	MISC. FCT.
Load	001	0	00000	00000	0	00	02		018	9	01300	02462	0	00	80
Tool change	002	0	00000	00000	0	01	86	3rd. hole	019	4	06500	08291	0	00	80
Side 1 1st. hole	003	4	06500	06850	0	29	88		020	9	01300	02350	0	00	80
	004	9	01300	02350	0	02	89	Retct. spin	021	0	06500	08291	0	00	82
2nd. hole	005	4	06500	08421	0	00	80	Index 45°	022	0	06500	07558	0	00	83
	006	9	01300	02128	0	00	80	Index 45°	024	0	06500	07558	0	00	83
Retct.	007	0	06500	08421	0	00	82	Side 3	025	4	06500	07558	0	00	80
Index 45°	008	0	06500	07426	0	00	83		026	9	00742	02104	0	00	89
Index 45°	009	4	06500	07426	0	00	83	Retct. spin	027	0	06500	07558	0	00	82
Side 2 1st. hole	010	9	00742	01745	0	02	89		028	0	00000	00000	0	00	80
2nd. hole	011	4	06500	07690	0	00	80	Index 45°	029	0	00000	00000	0	00	83
Retct.	012	9	00742	01745	0	00	82	Index 45°	030	0	00000	00000	0	00	83
Index 45°	013	0	06500	07690	0	00	83	Side 1 ready	031	0	00000	00000	0	00	86
Index 45°	014	0	06500	07690	0	00	83	Tap Side 1	032						
Side 3 1st. hole	015	4	06500	06695	0	02	89								
	016	9	01300	02128	0	02	89								
2nd. hole	017	4	07200	06500	0	00	80								

Figure 28

6

OPERATIONS IN THREE AXES

6.1. THE THREE-AXIS MACHINE

Two-axis machines can execute standard machining cycles in the third axis by calling up what are described as "canned cycles" or sequences. The machine positions in two axes and then executes the canned sequence in the third axis. The Cintimatic vertical spindle two-axis mill can tap in the third or Z axis by calling up the canned preparatory cycle G85 with a preset cam for Z depths. This cycle contains the following movements:

1. Advance the quill rapidly to the Z feed point, which has been preset manually.
2. Tap in at feed rate to the terminal Z position.
3. Reverse the spindle rotation.
4. Tap out at feed rate to the Z feed point.
5. Retract the quill rapidly.
6. Reverse the spindle again.

Here are six operations, four of them being Z movements, and two being spindle reversals. The operations also call for two feed rates. In the operation of a three-axis mill, there are usually no canned cycles, and each of the above six operations must be individually programmed.

There are almost as many manufacturers of three-axis mills as of two-axis machines. Cincinnati Milacron, for example, produces both types. Kearney & Trecker of Milwaukee, on the other hand, builds only machines in three or more axes.

Three-axis mills are available in horizontal and in vertical spindle configurations. A vertical spindle machine is the choice for die-sinking and other operations requiring work on one face of the part only. Fixturing is usually rudimentary in vertical spindle machine work, since the part can be clamped to the work table or held in a suitable milling vise. Chip accumulation, however, is an inconvenience.

The Gorton Tapemaster 2-30 of Fig. 29 is an example of a vertical

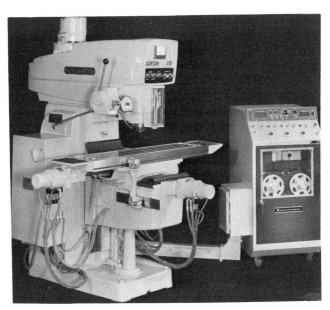

Fig. 29. The Gorton Tapemaster, a numerically controlled contouring vertical spindle milling machine. (Courtesy of George Gorton Machine Co.)

spindle three-axis mill. This machine is a contouring mill, also available as a two-axis machine.

For production work, or work requiring operations on more than one face of the part, as is necessary in the manufacture of fluid power pumps, valves, and manifolds, a horizontal spindle is preferred. In this configuration chip accumulation is a lesser problem. The increased versatility of the horizontal spindle machine will become apparent in the comments of this chapter. However, this is usually a more expensive machine configuration, and usually more extensive fixturing is needed.

Both horizontal and vertical spindle machines are available as positioning or contouring machines. Many offer a choice of machine control unit, and the tape format may be word address or tab sequential. Dimensioning is absolute for positioning control units and usually incremental for contouring units.

The horizontal spindle machine configuration is well adapted to receive additional equipment to extend its versatility. Automatic tool changers with tool magazines, rotating work tables, or fourth and fifth axes may be added. Pallet shuttle devices are also used, in particular with the Milwaukee-Matics about to be discussed.

Tool turrets [Fig. 18(d)] and tool magazines with automatic tool changers (Fig. 30) are, of course, sometimes available in vertical spindle machines, such as the Moog machining center. The use of tool changer and magazine does not reduce the time required to change tools as compared with a manual tool change. Justification for such accessories lies in other considerations. Automatic tool handling may be required as insur-

(a)

(b)

Fig. 30. (a) Milwaukee-Matic Model II. (b) Milwaukee-Matic Series E machining center. (Photos courtesy of Kearney & Trecker Corporation, Milwaukee)

ance against the use of the wrong tool in a machining sequence. More commonly, this facility allows one operator to tend two machines.

The tool magazine can store and handle a greater number of tools than a tool turret. No turret stores more than 10 tools, though the turret can index the next tool into position faster than an automatic tool changer can exchange tools. Most tool changers require five to seven seconds for an exchange.

If the machine is equipped with such devices as pallet shuttle, rotating work table, or tool magazine and changer, it is referred to as a *machining center,* meaning a machine tool capable of a wide range of machining operations in one setup, normally performed by a number of different types of standard machines.

Perhaps the best-known family of multi-axis machining centers, and perhaps also the most interesting for detailed consideration, is the group of Kearney & Trecker Milwaukee-Matics. These are all horizontal spindle types with automatic tool change and tool magazine, and an indexing table. Currently these are equipped with the following machine control units:

      Model Ea —Square D control
      Model Eb —General Electric
      Model II —Bendix, General Electric, or Allen Bradley
      Model IIIb—General Electric

Kearney & Trecker produce also the larger and more complex Models IV and V.

Depending on the machine control unit, the tape format for a Milwaukee-Matic may be either tab sequential or word address. The smaller machines, such as the two Model E types, may be equipped for positioning or positioning-contouring control. The larger multi-axis machines are contouring models. The contouring machines and those machines of the positioning type with a contouring mode are programmed by dimensional incrementing; the positioning machines use absolute dimensions. The zero point or origin of coordinates cannot be shifted.

The indexing tables on the smaller Milwaukee-Matics do not provide a fourth machining axis. The table does not index during machining operations, and is, therefore, a work-handling fixture rather than an additional axis.

6.2. PROGRAMMING THE MILWAUKEE-MATIC MODEL II

This machine is illustrated in Fig. 30. An eight-position indexing work table (a 360,000 angular position table is available as an option) and a pallet shuttle mechanism are provided. The machine can work on one part while the operator sets up another part on the second pallet. At the end of the tape sequence, the working pallet is shifted to the end of the machine bed and the other pallet is shuttled into position for machining. Since the machine control unit has two tape readers, the two parts set up on the shuttle pallets need not be identical; the tape for a different work piece can be placed in the second tape reader.

The tape format is word address if the machine is equipped with General Electric integrated circuit controls. If the tape format is tab sequential, each word is preceded by a TAB punch, except for the sequence number.

The programming of a positioning-control Model II will be first explained. Contouring operations will be programmed in Chapter 7.

A block of information for positioning control contains the following words and digits, not including sequence number.

Word		Digits
X		6 - 4 decimal places
Y		6 - 4 decimal places
Z		6 - 4 decimal places
F	Feed rate	6 - 3 decimal places
B	Table index	1
S	Spindle speed	4 - no decimal places
T	Tool code	4
M	Miscellaneous function	2

Thus with letter addresses and EOB, but not including sequence number, a complete block of information with all digits would be 44 characters or 4.4 in. long. However, full blocks of information are unusual in three-axis programming.

Trailing zeros may be omitted, except for tool code numbers and miscellaneous functions.

Feed Rate. This is coded in six digits. The decimal point is understood to follow the first three digits. Thus 010 indicates a feed rate of 10 ipm. Rapid traverse is coded as the single digit 8.

Eight-Position Table Index. The number of the table position as indicated in Fig. 31 is punched.

Spindle Speed. Speeds range from 1 to 4000 rpm in 1-rpm increments. Thus either 1 or 1000 indicates 1000 rpm.

Tool Code. Each toolholder has spaces for two groups of five coding rings each, or a total of 10 spaces. See Fig. 32. The five spaces at the shank end of the toolholder comprise the Tool Group number section, and the

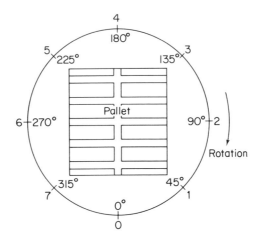

Figure 31

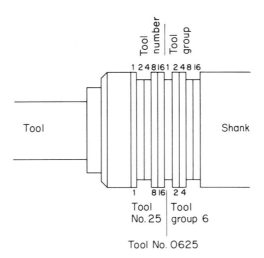

Figure 32

five spaces at the tool end of the holder comprise the tool number. Rings are placed in these spaces for coding. Tool groups 30 and 31 are reserved for taps.

Suppose that tool group 06 is coded, and a tool in this group is to be coded 25. Two coding rings placed in the 2 and 4 positions in the Tool Group section code the 06. Three coding rings in the Tool Number section, 1, 8, and 16, code tool number 25. Thus the tool is coded 0625. The binary system is thus used to code the spaces.

The standard miscellaneous function 06 calls for a tool change. Evidently the next tool for the tool change must be called up before a tool change is made. The tool number to be used next is programmed in some block of information preceding the block containing the 06 tool change code. This will be understood from the programming example to follow.

MISCELLANEOUS FUNCTIONS. These are standard functions as used with most n/c machines.

1. 00. A programmed cycle stop for any purpose. Tape reading is continued by depressing the Cycle Start button.
2. 01. Optional stop. This programmed stop will not occur unless the operator switches the optional stop selector switch on the control panel before this function is read.
3. 02. End of program.
4. 03. Clockwise spindle rotation, as viewed from the spindle.
5. 04. Counterclockwise spindle rotation.
6. 05. Spindle stop. The spindle will rotate at the last speed commanded unless a new spindle speed is entered before the spindle is restarted.
7. 06. Automatic tool transfer to spindle with removal of tool in spindle back to tool magazine.
8. 07. Mist coolant on. Coolant automatically turns off when Z axis retracts to home position and turns on when Z advances.
9. 08. Flood coolant on. Coolant turns off and on with Z motions as explained for the 07 function.
10. 09. Coolant off. This command is required when one is changing from 07 to 08 or 08 to 07.

6.3. COORDINATE AXES OF THE MODEL II

Because of the pallet shuttle facility and other reasons, no zero shift is possible. The origin is fixed. The coordinate system appears somewhat unusual on first acquaintance. It is illustrated in Fig. 33. Note that the origin of XY coordinates lies 1.0000 in. above the pallet surface, and that the Z axis is independent of the XY coordinate system. This unusual geometry is, of course, advantageous for programming movements of the spindle in and out.

The exact center of the surface plane of the work pallet has the coordinates

$$X = +12.0000$$
$$Y = -01.0000$$
$$Z = +20.7500$$

X is positive to the operator's left, Y is positive upward, and Z is positive toward the work from the spindle. There are no minus X, Y, or Z coordinates.

Interference between tool and work piece or fixture is possible when one is indexing the work on the rotating table. The table, therefore, must be indexed to a new rotational position when the spindle is retracted. As Fig. 33 shows, the Y dimension cannot be programmed to less than three inches when the Z axis is programmed beyond 11 inches. Otherwise, maximum travel of the Z axis is 16 in. The maximum Y dimension is 20 in. The maximum X dimension is 24 in.

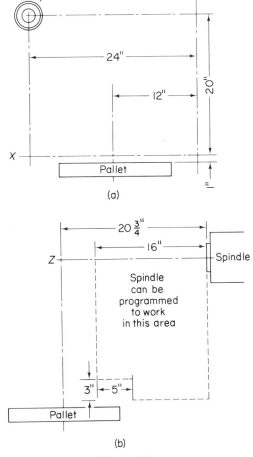

Figure 33

Figure 34 shows coordinate relationships for coordinates X130000 Y100000 Z150000. Note that Z coordinates are measured from the face of the spindle. If the spindle is retracted to Z000000 and the tool projects 05.0000 in., then the point of the tool is located at Z050000 when in the retracted position.

6.4. THE PART PROGRAM FOR POSITIONING CONTROL

The many small details set out above leave a first impression that three-axis programming, for a Milwaukee-Matic at any rate, may be somewhat difficult. Fortunately, this first impression turns out to be erroneous.

The work piece to be programmed here is the offset link, part print no. 4. There are some contouring operations to be performed on this part, but these will for the moment be ignored, and positioning opera-

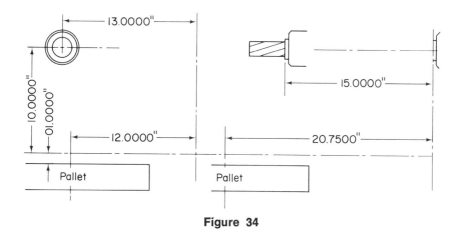

Figure 34

tions will first be discussed. The positioning operations include the drilling of six holes and the milling of surface B.

Four sides of the work piece must be presented to the spindle for machining operations. It is desirable to complete all machining in a single setup. For this to be done, the part must be mounted in the position shown in Fig. 35. Tooling for such a setup will be somewhat elaborate. Tooling details are not discussed, but the reader should be able to decide the tooling configuration required, so that all four sides can be successively indexed for machining. Presumably this tooling would be manufactured by the Milwaukee-Matic itself.

Programming of the two-axis machines discussed early in this book was relatively easy. XY coordinates were determined, and then preparatory and miscellaneous functions were attached to these coordinates. In a

84 *Chap. 6 / Operations in Three Axes*

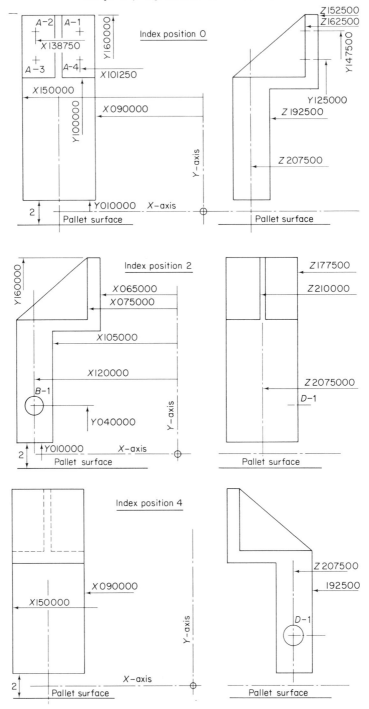

Figure 35

three-axis program, there is much more detail to control, so that the programmer must organize this detail stage by stage. Here the program is worked out by means of three manuscript sheets in order to keep control of the data: a tool sheet, a coordinate sheet, and the program sheet. By working through these three manuscript sheets, the required program data are compiled step by step in an organized manner.

The tool sheet, Fig. 36, accumulates the tool information, including tool length (setting distance), tool code number, spindle speeds, feeds,

TOOL SHEET

PART NUMBER AND NAME							
OPERATION		MATERIAL		PAGE OF			
NO.	OPERATION	TOOL	SETTING DISTANCE	TOOL CODE	RMP	IPM	INDEX POSN
1	Face mill side A	Face mill 2½"	4.000	0101	0300	5	0
2	Drill holes in A	Drill ¹⁵⁄₁₆"	6.000	0102	0200	3	0
3	Mill surface B	Mill 1½"	5.100	0103	0400	5	0
4	Core drill sides B,C	1⁷⁄₁₆"	7.000	0104	0180	3	2,6
5	Bore holes B,C		6.000	0105	0350	3	2,6
6	Ream holes B,C	Reamer 1½"	7.500	0106	0200	5	2,6
3	Mill stiffener C	Mill 1½"		0103			2
7	End mill circle	End mill 0.25	3.000	0107	0400	5	4

Figure 36

and table index position. The programmed Z distances can then be determined from the setting distances of the individual tools. In scrutinizing the tool sheet, note that tool items 3 and 7 are used for two contouring operations to be discussed; item 3 is a milling operation at an angle to the axes, and item 7 requires milling of a curve.

The work piece is set up on the pallet with surface A (with four bolt holes) facing the spindle, and the axis of the part as drawn collinear with the axis of the pallet. The indexing table is initially at the 0 deg index position. After the required machining in the zero table position, the table will be indexed clockwise to 2 position, presenting hole location B-1 to the spindle. The next index position will be 4, and finally 6. At the end of the machining program the table must be indexed back to zero position. The successive index positions are sketched in Fig. 35. The left-hand views are those seen by the operator of the machine, parallel to XY plane; the right-hand views are YZ views. These views assist in following the details of the coordinate sheet.

COORDINATE SHEET

PART NO. AND NAME									
HOLE CODE	X	Y	Z_C	Z_D	HOLE CODE	X	Y	Z_C	Z_D
A-1	101250	147500	151500	166500					
A-2	138750	147500	151500	166500					
A-3	138750	125000	151500	166500					
A-4	101250	125000	151500	166500					
B-1	120000	040000	176500	198750					
D-1	120000	040000	176500	198750					
A-5	090000	160000	—	163500					
A-6	150000	160000	—	163500					
B-2	075000	160000	—	211000					
B-3	133500	109000		211000					
C-1	15000	083750		193437					
C-2	126846	083750		193437					
C-3	120000	087500		193437					
C-4	113154	083750		193437					
C-5	090000	083750		193437					

Figure 37

The coordinate sheet lists the X, Y, and Z coordinates of points and holes on the work piece. These coordinate values are given for the table index position to which they apply. The Z coordinates are the actual coordinates on the work piece; before being punched into the tape these numbers must be corrected for the setting length of the tool. The Z_c column refers to the clearance position from which the tool advances at feed rate; an allowance of 0.100 in. from the part is used in determining Z_c. The Z final depth position Z_D must allow for the length of the drill point, which is 0.3 × diameter. The last seven points, from B-2 to C-5, are required for contouring operations discussed in the following chapter.

6.5. THE PROGRAM SHEET

The part program sheet, Fig. 38, begins with the first tool 0101 already in the spindle for milling surface A, the table already indexed to position 0, and the spindle retracted. Only the positioning operations are discussed here, and the columns Preparatory Command, Contouring Feed Rate, and Arc Center Offset do not for the moment apply, since they provide contouring information. If the machine is a positioning machine, these words would not be a part of the information block.

The Program Sheet

PROGRAM SHEET   MATERIAL   PAGE NO 1 OF 3

PART NUMBER AND NAME
OFFSET LINK

SEQ NO.	OPERATION	PREP COMM	X	Y	Z	FEED RATE	TABLE INDEX	SPIN SPEED	TOOL CODE	AUX FCN	CONT. FEED RATE	ARC CENTER OFFSET X/X/Y	Y/Z/Z	TIME
	Mill surface A and drill 4 holes in surface A													
1	Rapid to XY	00	14	086	0	8								
2	Spindle on							03		03				
3	Rapid Z				1125	8				08				
4	Mill up			174		005								
5	Rapid X		12			8								
6	Mill down			086		005								
7	Rapid X		10			8								
8	Mill up			174		005								
9	Retract Z				0	8								
10	Tool 0102								0102	05				
11	Rapid to A-2		13875	1475	0	8				06				
12	Spindle on						0	02		03				
13	Rapid Z				0915	8								
14	Drill A-2				1065	003								
15	Retract Z				0915	8								
16	Rapid to A-1		10125	1475	1065	8								
17	Drill A-1				0915	003								
18	Retract Z				1065	8								
19	Rapid to A-4			125		8								
20	Drill A-4				1065	003								
21	Retract Z				0915	8								
22	Rapid to A-3		13875	125		8								
23	Drill A-3				1065	003			0103					
24	Retract Z				0	8				05				

Figure 38

88   *Chap. 6 / Operations in Three Axes*

PROGRAM SHEET   MATERIAL   PAGE NO 2 OF 3

PART NUMBER AND NAME

SEQ NO.	OPERATION	PREP COMM	X	Y	Z	FEED RATE	TABLE INDEX	SPIN SPEED	TOOL CODE	AUX FCN	CONT FEED RATE	ARC CENTER OFFSET X/X/Y	OFFSET Y/Z/Z	TIME
25	Change, tool 0103									06				
26	Rapid, XY		159	1675	0	8								
27	Spindle on							03		03				
28	Rapid Z				1125									
29	Mill surf B		81		0									
30	Retract Z									05				
37	Index to 2						2							
38	Rapid to B-2		078018	16663	0	8								
39	Spindle on									03				
40	Rapid Z				16	8								
41	Dwell XY	44	01											
42	Mill slope C	10	0585	-051	0	007			0104	05				
43	Retract Z	00				8				06				
44	Change, tool 0104													
45	Rapid XY		12	04		8								
46	Rapid Z				1065	8		018		03				
47	Drill D-1				12875	003								
48	Retract Z				0	8								
49	Index to 6						6							
50	Rapid Z				1065	8								
51	Drill B-1				12875	003			0107	05				
52	Retract Z				0	8								
53	Index to 4						4							
54	Change tool, 0107									06				

Figure 38 (cont.)

The Program Sheet

PROGRAM SHEET

PART NUMBER AND NAME													PAGE NO 3 OF 3		
OFFSET LINK						MATERIAL									
SEQ NO.	OPERATION	PREP COMM	X	Y	Z	FEED RATE	TABLE INDEX	SPIN SPEED	TOOL CODE	AUX FCN	CONT FEED RATE	ARC CENTER OFFSET			TIME
												X/X/Y	Y/Z/Z		
55	Rapid toward C−1		15225	08375	0	8	4								
56	Rapid Z		126846		193438	8		04		03					
57	Mill C−1 to C−2		01			005									
58	Dwell XY	44													
59	Mill C−2 to C−3	21	−006846	00375		06						006846	004375		
60	Mill C−3 to C−4	21	−006846	−00375		06						00	008125		
61	Mill C−4 to C−5	00	08775			005									
62	Retract Z				0	8			0101	05					
63	Index to 0						1								
64	Change tool, 0101									06					
65	Shuttle pallet									02					

Figure 38 (cont.)

The following comments on the information blocks are offered for clarification.

1. The feed rate is programmed as 8 for rapid traverse in X, Y, or Z.
2. Auxiliary function 03 calls for clockwise spindle rotation at 300 rpm.
3. The spindle is turned on at 300 rpm.
4. The tool moves rapidly to the position of full depth of milling cut.
5. The first milling cut is made with a movement in Y up the part. In sequences 6 through 8 two more milling cuts are made to complete the required face milling.
9. The tool is withdrawn to the retracted position in preparation for a tool change.
13, 14, 15. To drill a hole requires three operational sequences: rapid advance to the feed point, slow feed to depth, and rapid retract.

At this point the general method of programming a three-axis machine should be apparent, as compared to programming a two-axis machine. A two-axis machine may have an information block of five words. The block for such a machine calls up an XY position, and the G word may call up a complex series of tool motions. This contouring three-axis Milwaukee-Matic uses an information block of 12 words, but as few as one or perhaps two words may be used in any block, because only one machine operation is programmed in any block. By comparison, an information block for a Cintimatic two-axis positioning mill can call up as many as seven machine operations, because of the use of canned cycles. It should be apparent that a three-axis machine is not more difficult to program.

41, 42. These two sequences contain contouring operations, which are discussed in the next chapter.
48. Note that the spindle must always be retracted before the table is indexed.
58–61. This contouring operation is explained in the next chapter.
63, 64. In preparation for producing the next offset link in the production run, the table is indexed back to the 0 position for sequence no. 1 in the program, and tool 0101 is placed in the spindle by the automatic tool changer.
65. The auxiliary function 02 (end of program) also shuttles the finished part to the unloading position at the end of the machine bed, and shuttles the other pallet into operating position.

The tape format for such a machine may be either tab sequential or word address. If tab sequential, a TAB code must be used as discussed in Sec. 2.8 to ensure that each word in the block is switched to the correct memory. A TAB code, however, does not precede the sequence number nor the preparatory command. To illustrate the tab sequential method, consider sequences 63, 64, and 65 of the program sheet. In tab sequential, these must be coded

```
63 T T T T T1       (EOB)
64 T T T T T T T T06 (EOB)
65 T T T T T T T T02 (EOB)
```

No TAB codes are required after the last piece of information in the block.

If this machine were a positioning machine only, the information block would not include the preparatory command, the contouring feed rate, nor the two arc center offset words.

Questions

1. Program the base pad, part print no. 2, for production on the Milwaukee-Matic Model II.
   The table remains indexed at zero position throughout the operation. The base pad has already been milled on all six sides, and requires only pocket milling and drilling. Omit the tapping operation in the small pocket.
   Affix the work piece in a picture-frame fixture, so that:
   a. The bottom of the base pad is located at Y050000.
   b. The vertical axis of symmetry of the base pad is located at X120000.
   c. The face of the base pad is located at Z160000.
   Assume all drills to have a setting length of 4.0000 in. and all mills 3.5000 in.
   In milling the small pocket, use two roughing passes only.
   In milling the large pocket, clean out the center of the pocket with a 5 in. mill, then remove the periphery of the pocket with a 1.000 in. mill.
   If uncertain of speeds and feeds, code these as 1's (111, etc.).

2. Figure 6-1 is a front plate for an air cylinder, and requires machining operations on three sides. These are to be completed in a single setup on the Milwaukee-Matic Model II.
   Affix the work piece so that
   a. The base is located at Y050000.
   b. The vertical axis of the part is located at X120000.
   c. The front view shown in the figure is located at Z187500.
   Assume all tools to have a setting length of 6.0000 in.

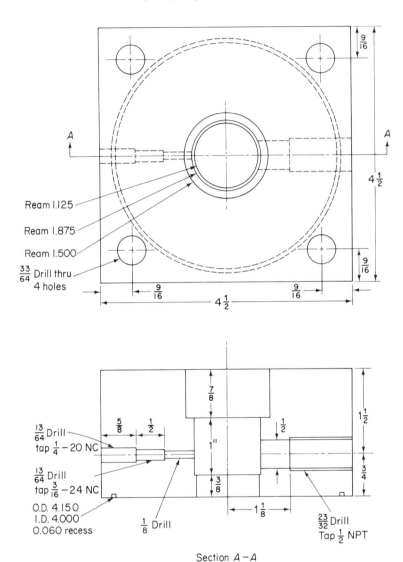

Section A−A

Figure 6-1

7

BASIC PRINCIPLES OF CONTOURING

7.1. CONTOURING COMPUTATIONS

The method of contouring circular arcs in stair-step fashion on a positioning mill as explained in Chapter 4 is not especially difficult, but is repetitious. In this chapter some of the principles and methods of contouring and contouring machines are discussed, and some examples are worked by hand calculation.

These contouring principles must be understood if programs are to be put into a computer with any hope of success. Sometimes, too, hand calculations for contouring must be undertaken by the programmer, though only for simpler cases. Most important of all reasons, the creativity needed to exploit the potential of numerical control can be realized only if contouring methods are understood.

When contouring computations become at all cumbersome or repetitious, then a computer becomes economical. The most complex programming discussed or suggested in this chapter can be executed on a computer for a few dollars. However, since a computer program frequently must be modified and rerun through the computer a second time, let us say that the computer cost of such programs is of the order of $10. If a programmer's time is worth $5 an hour, then any problem requiring two hours of computation or more must be given to the computer to solve. That two-hour limit must include checking of calculations also, for in hand programming we assume that any calculation can be erroneous. The examples in this chapter point to the conclusion that most contouring calculations of any complexity must be executed by a computer.

Any n/c machine with contouring capability, such as the Gorton Tapemaster mill, the Kearney & Trecker Milwaukee-Matic Model II, or the Monarch turN/Center 75, of the next chapter, is not restricted to straight lines parallel to X and Y axes, but can cut a straight line at any angle. If the machine is required to approximate a complex curve within a specified tolerance, the machine may do so by any of three possible methods:

1. By straight-line segments at any angle, or linear interpolation.
2. By circular arcs, or circular interpolation.
3. By parabolas, or parabolic interpolation.

Most contouring machines have only the capacity to cut straight-line segments, though lathes are equipped for circular interpolation. A limited number of larger machines use parabolic interpolation. Circular and parabolic approximation methods require more elaborate computing capacity in the machine control unit.

If the n/c machine is to approximate a curve by linear interpolation, then each approximating straight-line cut occupies a block of information. In the case of circular interpolation, the whole arc is programmed in a single information block with i, j, and k arc center offsets, which will be explained presently.

7.2. CONTOURING BY LINEAR INTERPOLATION

If a contouring mill must produce the O-ring groove of Chapter 4 by linear interpolation, then three possible types of straight-line segments may be used:

1. Chords (negative tolerance).
2. Tangents (positive tolerance).
3. Secants (positive and negative tolerance).

These three methods are illustrated in Fig. 39.

The arcs of the O-ring groove are to be approximated with a maximum error of 0.003 in. This tolerance band may mean either of two things: a mathematical error, or the sum of mathematical and machine errors. If we are to allow 0.001, say, for machine errors, then the mathematical tolerance must be restricted to 0.002.

Turning first to the use of chords, in Fig. 40 the approximating chord is AC, with midpoint D. Distance BD is the maximum allowable error n, taken as 0.003, say. From the figure

Contouring by Linear Interpolation

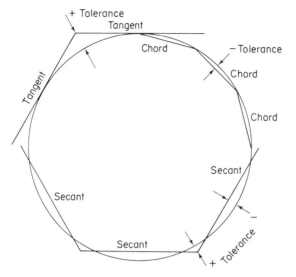

Figure 39

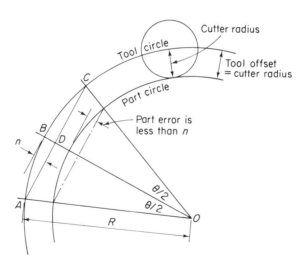

Figure 40

$$n = R - (OD)$$
$$(OD) = R - n$$

where R is the sum of the part radius plus the cutter radius, if the tool is offset from the part.

Also,
$$(AO)^2 = (OD)^2 + (AD)^2$$
which is
$$R^2 = (R - n)^2 + (AC/2)^2$$
so that
$$(AC)^2 = 4[R^2 - (R - n)^2]$$
$$= 4[2Rn - n^2]$$

But for all practical purposes in numerical control, n^2 can be neglected, being far less than the least positioning increment of the machine. Finally, therefore,

$$(AC) = 2\sqrt{2Rn}$$

This is the maximum chord length for the allowable tolerance. For the O-ring groove arc of radius 0.641 ± 0.003, the chord length is 0.124 in.

To determine the number of chords required, the angle θ is found from the formula

$$\theta = 2 \cos^{-1}\left[\frac{R - n}{R}\right]$$

Again for the 0.641-in. arc, ±0.003, θ is 11°6'.

The arc under examination covers 120°, so that a chord equivalent to 11°6' is not convenient, since it would make for a special calculation for the last chord. A chord length equivalent to 10° would be selected, and the arc approximated by exactly 12 equal chords.

In Fig. 40 the tool path radius is larger than the part radius. Therefore, the error in cutting the part circle is less than the error n for the tool circle. The larger the tool diameter, the smaller will be the part circle approximating error. The ratio between the two errors is not, however, the ratio of the part diameters. If, however, the tool must cut inside the part circle, then the part error will exceed the tool circle error. Hence, if one is cutting outside the part circle, a large-diameter tool will produce a smaller part error, and if one is cutting inside the part circle, a small-diameter tool will reduce the part error.

Methods of straight-line approximation, such as the chordal method, which use negative tolerance, are not recommended for the cutting of small-diameter circles.

If the arc of 120° and radius 0.641 in. must be programmed by hand calculation, then 12 sets of X and Y coordinates must be calculated. This is not too difficult a job with a desk calculator. The work may also be done with a computer programmed in Fortran language. Twelve chords to approximate the arc are a fraction of the number of increments required when the machine can move only in X and Y, as in Chapter 4.

For small diameters the tangent method is preferred, because the error lies outside the part circle. In this method half-tangents are used for the first and last span, as seen in Fig. 41. The geometry of the tangent method is similar to that of the chord method.

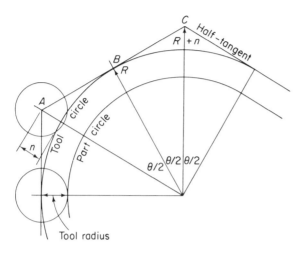

Figure 41

$$(R + n)^2 = (BC)^2 + R^2$$

Neglecting n^2 as before, we find that this reduces to

$$2Rn = (BC)^2$$
$$BC = \sqrt{2Rn}$$
$$AC = 2\sqrt{2Rn}$$

This is the same result as for chords; therefore, the number of tangent spans to approximate an arc is the same as the number of chord spans:

$$\text{Number of spans} = \text{total angle}/\theta$$
$$\frac{\theta}{2} = \cos^{-1}\frac{R}{R + n}$$

The use of secant approximation provides longer spans than are given by tangents or chords, because a bilateral error is accepted. Commonly the first and the last span in secant approximation is made a half-tangent, so that the beginning and the end point of the curve-fitting lie on the desired curve.

The secant geometry is given in Fig. 42.

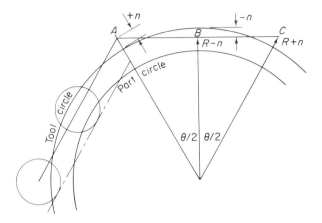

Figure 42

$$(BC)^2 + (R-n)^2 = (R+n)^2$$

Since n^2 is negligible, this relationship reduces to

$$(BC)^2 - 2Rn = 2Rn$$
$$(BC)^2 = 4Rn$$
$$BC = 2\sqrt{Rn}$$
so that AC = secant = $4\sqrt{Rn}$

When secants are used, only two-thirds as many straight-line segments are required as compared with tangents or chords.

$$\text{Number of spans} = \text{total angle}/\theta$$
$$\theta = 2\cos^{-1}\frac{R-n}{R+n}$$

For the secant case, if $R = 0.641 \pm 0.003$ over $120°$, then $\theta = 15°40'$. This would be adjusted to $15°$, thus approximating the arc with eight segments.

Questions

1. Figure 7-1 shows a short length of part contour of 12° of arc, beginning at point A and terminating at point F. The part contour is to be machined with an end mill of diameter 0.500 using a tolerance of ± 0.002. The first and last straight-line segments are to be half-tangents.

    Program the increments in X and Y between points on the tool path (incremental dimensions are used in contouring). If at all possible, use a desk calculator, since more than three decimal places are required.

2. The cosine of the increment angle in the above question is 0.998xx. Suppose this cosine were calculated to only three decimal places, that is, 0.998. What angular range does a cosine of 0.998 cover?

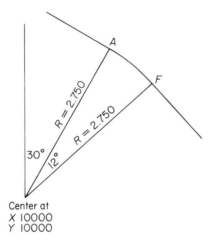

Figure 7-1

7.3. CIRCULAR INTERPOLATION

An n/c machine equipped for circular interpolation can program an arc in a single block of information. In programming circular interpolation, the machine control unit requires the following data:

1. The coordinates of the start of the arc.
2. The coordinates of the end of the arc.
3. The arc-center offset coordinates I, J, K.

The tool is directed to the starting point of the arc by the previous motion command on the tape, so that this previous command contains the coordinates of the start of the arc. The arc command must, of course, supply the coordinates of the end of the arc.

The arc-center offsets are the distances from the arc center point to the start of the arc, as shown in Fig. 43. The X distance from arc center to arc start is coded on the tape as the I word, the Y distance as the J word, and the Z distance as the K word. When plus and minus signs are required in programming dimensions, no sign is used with the I, J, or K words.

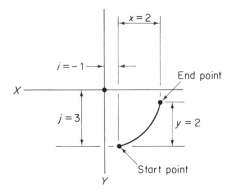

Figure 43

Almost all n/c machines are limited to a maximum arc of 90 deg in one quadrant per block of information. Coding of a full circle, therefore, would require four information blocks.

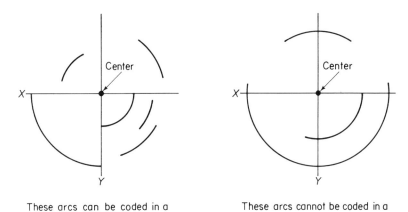

These arcs can be coded in a single information block

These arcs cannot be coded in a single information block

Figure 44

Examples of I, J, and K words will appear in the programming examples of this and the following chapter. Note that $R = \sqrt{I^2 + J^2}$.

Special G functions are commonly used to call up linear and circular interpolation. These functions will be discussed in the following chapter as applied to lathe turning. Although the following are standard, some machines employ different contouring preparatory functions.

    G01—linear interpolation. To be coded for any straight line, including lines parallel to any axis.

    G02—circular interpolation, when the circular arc is cut clockwise.

G03—circular interpolation, when the circular arc is cut counterclockwise.

The interpretation of the terms clockwise and counterclockwise is best defined in terms of each specific machine, since there are always exceptions to general rules.

7.4. PARABOLIC INTERPOLATION

The familiar French curves used in drafting practice are parabolic curves, for the reason that the best curve-fitting shape is the parabola. For the same error or tolerance, fewer parabolic segments than straight lines are required to approximate any curve, as Fig. 45 shows. Further, any two

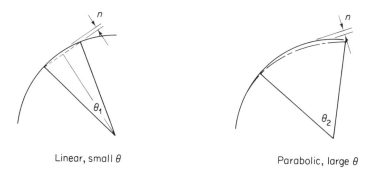

Figure 45

adjacent straight lines meet at a small angle, resulting in a poor surface finish. Ideally, the approximating lines should be tangent to each other, and this is possible with the use of parabolas. (It is also possible with approximating circles, of course.)

Two possible parabolas may be used for curve-fitting. One is the tangent parabola, the other the three-point parabola. Both possibilities are illustrated in Fig. 46.

The tangent method is used in drafting practice for curve-fitting by eye. In this method, the adjacent parabolic segments are tangent to each other. The maximum error occurs at the midpoint of the span of the approximating parabola, as the figure shows. In the three-point case, the parabola lies exactly on the actual curve at the end points and the midpoint of the parabola. The maximum error is between pairs of these three points. This latter parabola produces a smaller approximating error than the tangent parabola, but is not tangent to the desired curve at its ends. Therefore, fewer parabolic segments are required to approximate a curve to a

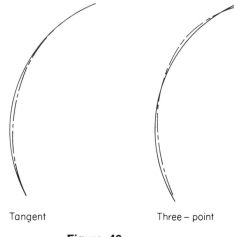

Figure 46

given error by the three-point method, but the lack of tangency is possibly inconvenient.

The manual methods of computation for parabolic approximation of curves are complex indeed and too time-consuming to merit discussion here. Those n/c machines with parabolic interpolation can compute the required parabola if given three points and the allowable tolerance as input information.

7.5. CURVE-FITTING

Sometimes curves must be defined from an empirical set of points. Special cases of such curve-fitting may be relatively simple in two-dimensional space, but most cases will call for computer approaches. The following example is a simpler case. The curve-fitting calculations are worked, but the tool offset calculations cannot be undertaken here. These require the use of an APT computer program. The example will give the reader some insight into the programming methods of the APT system, despite the fact that in an APT solution the mathematical method is slightly different.

In Fig. 47 a complex curve is shown, defined only by the XY coordinates of four points which must lie on the curve. Any smooth curve that fits the four given points will be acceptable.

The part programmer must make some strategic decisions before he undertakes any mathematical analysis. His first decision should perhaps be this: the smaller and more convenient the numbers to be manipulated, the faster the job is done and the fewer the sources of error in computation. The X and Y values of the four points are too large. For ease of computation make $X_1 = 0$ and $Y_1 = 0.1$, that is, reduce all X's by

Curve-Fitting

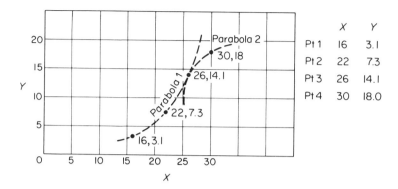

Figure 47

16.0 and all Y's by 3.0, thus temporarily displacing the coordinate system. Afterwards the required coordinate system can be restored. The following altered values result:

$$X_1 = 0 \qquad Y_1 = 0.1$$
$$X_2 = 6 \qquad Y_2 = 4.3$$
$$X_3 = 10 \qquad Y_3 = 11.1$$
$$X_4 = 14 \qquad Y_4 = 15$$

Reducing the Y values by 3.1 would be equally acceptable.

For curve-fitting two parabolas are selected. A parabola passing through P_1, P_2, P_3 has a vertical axis, opens upward, and has the general equation (from any suitable textbook on basic mathematics)

$$y = a_1 x^2 + b_1 x + c_1$$

A second parabola will be passed through P_3 and P_4. This could be a parabola with a vertical axis and opening downward, with an equation similar to that of the first parabola selected, or one with a horizontal axis and opening to the right. The two possibilities will, of course, produce two slightly different curves between P_3 and P_4. Let us suppose that gambler's choice is a parabola opening to the right, horizontal axis. Then its equation is

$$x = a_2 y^2 + b_2 y + c_2$$

(In contouring work the part programmer needs some familiarity with analytical geometry. A convenient text on this subject is the small *Analytic Geometry,* Barnes & Noble College Outline Series No. 68, priced at $1.75.)

For this second parabola there are three unknown coefficients,

a_2, b_2, c_2. We have the coordinates of two points P_3 and P_4 for input information. But three unknowns require three input items. The third item is the slope at P_3. The two parabolas must be tangent to each other at P_3, and this slope will be known when the first parabola is solved.

The first parabola, through P_1, P_2, P_3:

$$y = a_1 x^2 + b_1 x + c_1$$

Substitute the X and Y values of the three points into this equation to obtain three simultaneous equations:

$$0.1 = 0 + 0 + c_1$$
$$4.3 = 36a_1 + 6b_1 + c_1$$
$$11.1 = 100a_1 + 10b_1 + c_1$$

Then from the first of these equations $c_1 = 0.1$. Use this value in the other two equations to give

$$4.2 = 36a_1 + 6b_1$$
$$11 = 100a_1 + 10b_1$$

Eliminate b_1 by multiplying the first equation by $\frac{10}{6}$ and subtracting one equation from the other. Then $a_1 = 0.1$, and finally $b_1 = 0.1$.

The equation of the first parabola is

$$y = 0.1x^2 + 0.1x + 0.1$$

The decimal fractions could be removed by multiplying both sides of the equation by 10. But 0.1 is not an inconvenient number to manipulate. Leave the equation as it stands.

We require the slope of the parabola at P_3 in order to solve the second parabola, which must have the same slope. The equation for the slope of a parobola at any point may be obtained from mathematics texts, or by simple calculus, which is done here.

Slope at $P_3 = m = dy/dx = 0.2x + 0.1$, with $x = 10$, giving a slope of 2.1.

The second parabola has the equation

$$x = a_2 y^2 + b_2 y + c_2$$

Substitute the coordinates of P_3 and P_4 to obtain two simultaneous equations:

$$10 = 123.21a_2 + 11.1b_2 + c_2$$
$$14 = 225a_2 + 15b_2 + c_2$$

For the third simultaneous equation, use the slope at P_3.

$$\text{Slope} = m = \frac{dy}{dx}$$

But $dx/dy = 2a_2 y + b_2$ for the second parabola.

$$\frac{dx}{dy} = \frac{1}{2.1} = 0.47619 = 2a_2 y + b_2 \qquad \text{with } y = 11.1$$

The three simultaneous equations for the second parabola are then

$$22.2a_2 + b_2 = 0.47619$$
$$10 = 123.21a_2 + 11.1b_2 + c_2$$
$$14 = 225a_2 + 15b_2 + c_2$$

Eliminating c_2 between the latter two equations, we have

$$4 = 101.79a_2 + 3.9b_2$$

There are now two simultaneous equations in a_2 and b_2. The computations give

$$a_2 = 0.140881$$
$$b_2 = -2.65137$$
$$c_2 = 22.0727$$

The equation of the second parabola is

$$x = 0.140881y^2 - 2.65137y + 22.0727$$

This solves the problem of fitting a suitable curve to the given four points. The choice of curves was, of course, arbitrary.

This example is an artificial one, but develops the procedure for laying out contours. The methods of contouring can now be applied to some real part, such as the centrifugal fan scroll casing of part print no. 5, in preparation for the oxyacetylene flame-cutting of this shape with an n/c cutting machine.

Examination of the fan scroll shape, however, discloses some formidable computational difficulties. If this spiral shape is to be fitted by a series of tangent parabolas, these parabolas do not have axes parallel to the X or Y axis, but inclined to them. The equations for such parabolas will be of the form

$$ax^2 + bxy + cy^2 + dx + ey + f = 0$$

Six unknown coefficients require six simultaneous equations, to be solved for every one of the fitting parabolas. Any practical solution to such a problem must use the computer methods of the following chapters. After the curve-fitting is completed, there remains the more onerous problem of the X and Y components of the tool offsets, which continuously vary around the spiral.

Finally, in three-dimensional space, surfaces that can be defined by a mathematical function usually may be represented by the general quadric function

$$ax^2 + by^2 + cz^2 + \frac{fyz}{2} + \frac{gxz}{2} + \frac{hxy}{2} + \frac{px}{2} + \frac{qy}{2} + \frac{rz}{2} + d = 0$$

For example, a circle in the XY plane with a radius of 1 and center at $X = 0$, $Y = 0$ would have the following coefficients if represented as a general quadric:

$$a = b = 1, \quad d = -1, \quad c = f = g = h = p = q = r = 0$$

A straight line would have all coefficients zero except p, q, and d. The line $X = 3$ would have zero coefficients except for $p = 1$ and $d = -3$. The two parabolas in the curve-fitting problem just discussed could also be represented as a general quadric.

The general quadric equation is one of the more complex mathematical functions used in n/c contouring. It can be processed as a standard routine by the APT computer system discussed in later chapters.

7.6. TOOL OFFSET CALCULATIONS

The most troublesome work in numerical control is undoubtedly the manual calculation of points on the tool path for cutting a contour. Manual calculations are made only for simple contours of lines and arcs; a computer must be used for more complex shapes.

There are many problem cases in tool offset calculations and a variety of methods for solving them. Many are mathematicians' delights but too time-consuming and with too much potential for mistakes. The method of attack will be illustrated by the contour of Fig. 50, and other examples are given in the following chapter.

The basic problem is the calculations for X and Y at a corner point, as shown in Fig. 48(a). The tool is offset from the part by its radius. In the figure it is cutting vertically upward, then turns through an angle θ to make the corner point B and shape the straight line BC.

Tool Offset Calculations

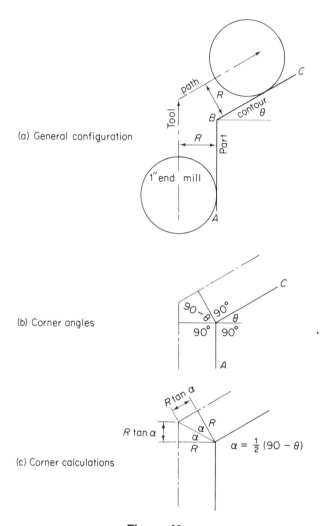

(a) General configuration

(b) Corner angles

(c) Corner calculations

Figure 48

The corner angles are indicated by Fig. 48(b). Two lines are erected from point B, one at right angles to the tool path that contours AB and the other at right angles to the tool path contouring BC. At the corner B there are three 90-deg angles and the part contour angle θ; therefore, the angle at B must be $(90° - \theta)$, since all angles must sum to 360 deg. This angle at B is designated as two angles α in Fig. 48(c). The two triangles so formed at B in Fig. 48(c) are identical, since they have equal sides and a right angle each.

In Fig. 49 the problem is a corner with two angular lines. The corner point where the two lines of the tool path intersect must be determined. This case may be solved by trigonometry or by analytical geometry. Figure 49 is a solution by analytical geometry.

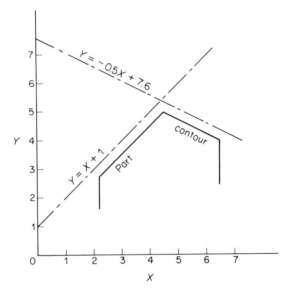

Figure 49

The equation of a straight line is given by

$$y = mx + b$$

where m is the slope or the tangent to the X axis. The slope is positive for a line with an angle counterclockwise from the X axis. Also b is the Y coordinate of the intersection of the line and the Y axis.

The slope of a line on the cutter path is that of the corresponding line on the part. The intercept b is determined as that point (Y coordinate) of the line where $X = 0$.

The two lines in Fig. 49 have angles with the X axis of $+45°$ and $-26°34'$ and therefore slopes of $+1$ and -0.5. Their equations are

$$y = x + 1$$
$$y = -0.5x + 7.6$$

At their intersection point both lines have the same X and Y coordinates. Solve the two equations as simultaneous equations. The result is, for the intersection point,

$$X = 4.4$$
$$Y = 5.4$$

The part contour of Fig. 50 is presented as an example of the required calculations for a tool path. We assume that the n/c machine has circular interpolation, so that circle C1 need not be approximated. The part contour is a simple shape with few dimensions. The scarcity of dimensions, however, means more work for the programmer. Only simple trigonometry is required, however.

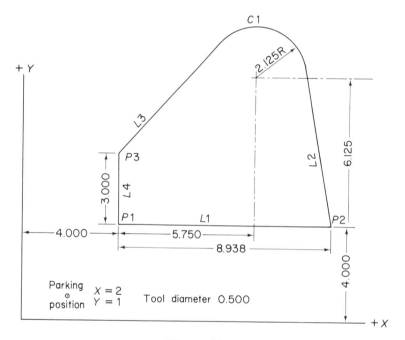

Figure 50

The tool, an end mill 0.500 diameter, is initially located at the parking position $X = 2$, $Y = 1$ as shown, and is to machine the contour in the customary counterclockwise direction. A clockwise machining sequence is, of course, entirely acceptable. The tool will be offset by 0.250.

The tool will begin cutting near P1, and will feed along the line $Y = 3.750$ to the region of P2. The turning point of the tool near P2 is not immediately known because line L2 has an unknown slope.

If we are to confine the tool path analysis to simple mathematics, we cannot calculate the tool path until we first find such data as the point of tangency of L2 with circle C1, C1 with L3, and other coordinates of the part contour itself. So we turn first to the part contour.

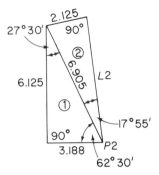

Figure 51

Figure 51 shows the two right triangles to be solved in series for data pertaining to L2. From right triangle 1 the distance from P2 to the center of C1 is determined as 6.905. The angles in triangle 1 must also be determined. These are given in the figure. In triangle 2, the side 2.125 is known and the side 6.905 has been calculated. The length of L2 can then be determined and also its slope. The length of L2 determines its point of tangency with C1.

L2 is found to be 6.573 in. long to the point of tangency with C1, and its slope from the vertical is 9°35′. The coordinates of the point of tangency are X11844 Y10481.

Exactly the same sequence of trigonometric calculations must be made for the point of tangency of L3 and C1 (Fig. 52). The coordinates

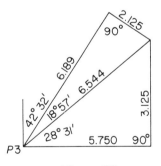

Figure 52

of all points on the part contour are given in Fig. 53.

Now for the points on the tool path. The tool starts cutting on the line $Y = 3.750$ near P1. Let us suppose that it can begin cutting at $X = 3.500$, at this point being off the unmachined contour, which perhaps is already rough-sawed. The calculations for the X coordinate of P2T are sketched in Fig. 55.

Tool Offset Calculations

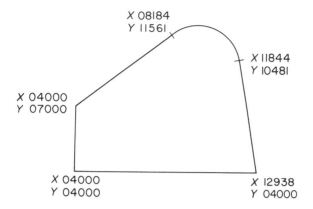

Figure 53

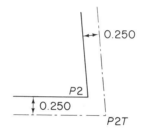

Figure 54

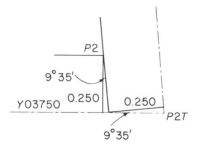

Figure 55

This X coordinate, by the methods of either Fig. 48(c), or Fig. 55,

$$= 12.938 + 0.250 \tan 9°35' + \frac{0.250}{\cos 9°35'}$$
$$= 13.234$$

For the position of the point of tangency of $L2$, $PC1L2$, see Fig. 56. The X coordinate of $PC1L2$

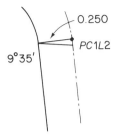

Figure 56

$$= 11.844 + 0.250 \cos 9°35' = 12.0905$$

and the Y coordinate

$$= 10.481 + 0.250 \sin 9°35' = 10.523$$

For the position of the point of tangency of $L3$, $PC1L3$, see Fig. 57.

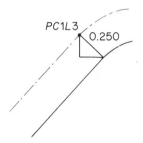

Figure 57

The X and Y coordinates for this point on the tool path are found from the cosine and sine components of the tool offset, as usual. The X coordinate of $PC1L3$

$$= 8.184 - 0.184 = 8.000$$

and the Y coordinate

$$= 11.561 + 0.169 = 11.730$$

for $P3T$ on the tool path near $P3$,

$$X = 4.000 - \text{tool offset} = 3.750$$
$$Y = 7.000 + 0.250 \sin 42°32' = 7.169$$

Such computational work as this is not inspiring, is time-consuming, and is certainly error-prone. Sometimes it must be done. But the preferred methods are the computerized methods of Part III of this book.

The X and Y positions and their increments may now be tabulated.

	X	Y	ΔX	ΔY
Parking Pos.	02000	01000	$+01500$	$+02750$
$P1T$	03500	03750	$+09734$	0
$P2T$	13234	03750	-01143	$+06773$
$PC1L2$	12091	10523	-02341	$+01977$
Crown	09750	12500	-01750	-00770
$PC1L3$	08000	11730	-04250	-04561
$P3T$	03750	07169	0	-03419
$P1T$	03750	03750	-01750	-02750
Parking Pos.	02000	01000		

Figure 58 compiles this data in a typical contouring program. The

ABSOLUTE		SEQ NO.	PREP FCN	DEPARTURES		ARC CENTRE OFFSET		MISC FCN	REMARKS
X	Y			X	Y	I	J		
02000	01000	001		000	000				Parking pos.
03500	03750	002		+01500	+02750				P1T
13234	03750	003		+09734	00000				P2T
12091	10523	004		−01143	−06773				PC1L2
09750	12500	005		−02341	+01977	02341	00398		Crown of arc
08000	11730	006		−01750	−00770	00000	02375		PC1L3
03750	07169	007		−04250	−04561				P3T
03750	03750	008		00000	−03419				P1T
02000	01000	009		−01750	−02750			02	Parking pos.

Figure 58

program format is a general one for circular interpolation, since a specific machine will have its own preparatory and miscellaneous functions and other special format items. Since this is a contouring operation, the dimensioning is incremental, starting from the parking position as X00000 Y00000. The increments are usually referred to as departures.

Since increments may be made in either of two directions, a plus and a minus direction, a sign convention must be established. Either the work table or the tool may be moved; the standard convention is to assume that the table is fixed and that the tool moves, even though the actual operation may be the reverse of this. If the tool moves to the right, this direction is taken as $+X$, while a tool movement to the left is in the $-X$

direction. If the tool moves away from the operator (up the page for Fig. 50), such a movement is in the $+Y$ direction; a movement toward the bottom of the page is a $-Y$ movement.

In contouring are $C1$ of Fig. 50, recall that only one quadrant of a circle may be coded in a block of information. Therefore, two blocks are required for this arc; the first block provides circular interpolation to the crown of the arc, and the second block to the point of tangency $PC1L3$. Recall also that I and J are the distances from the arc center to the start of the arc for every block with circular interpolation. I and J do not require a sign convention.

If the tool is returned to the starting point of the program, the sum of the positive and negative increments in either X or Y must total to zero. Figure 59 shows the standard G functions for circular interpolation in any of three planes.

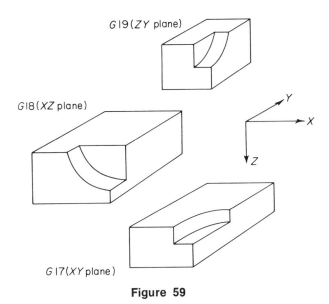

Figure 59

Questions

1. Figure 7-2 shows a template to be produced on a contouring mill with circular interpolation. Write an n/c program for this part with the manuscript format of Fig. 58. The tool diameter is 0.500.
2. Determine the required points on the tool path of Fig. 7-3. The tool diameter is 0.500. Counterclockwise machining around the part is sug-

gested. Write the manuscript for producing this part on a contouring machine with linear interpolation only.

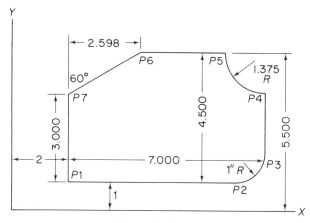

Figure 7-2

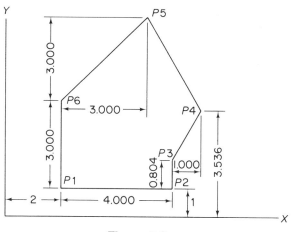

Figure 7-3

7.7. TAPE READING SPEED AND BLOCK PROCESS TIME

In contouring operations, especially during linear interpolation of a curved cut, very short movements of the tool may be employed. A problem exists here, in that the tape reader may not be able to read information into the n/c machine as fast as the machine may execute the information. One solution is the use of buffer storage; while the machine is executing the information in active storage, the tape reader is reading the next block into buffer storage.

Consider the milling of the O-ring groove on a positioning mill. This machine is not a contouring machine, but the problem is the same one. Suppose the groove to be milled at 3 ipm, this slow rate being dictated by the small size of the end mill. Now 3 ipm is 0.050″/second.

Suppose the information block for the XY stair-step contouring is typically some data such as N103X04223Y11897(EOB), a total of 17 characters. If the Cintimatic mill is used, this machine has a reading speed of 60 characters per second. Then the reading speed for this block of information is $\frac{09}{17}$ second, plus the time to start and stop the tape, which will be of the order of 0.01 second. Then reading time per information block is 0.293 second.

Compare this tape reading time with the operation time. If feed rate is 3 ipm, 0.050″/second, then in 0.293 second the tool will move 0.015″. For any cut shorter than 0.015″, the tool must wait for the next information block. Since the spindle will be rotating the tool as it waits, in effect a dwell will result, with loss of surface finish. When such short contouring cuts occur, the programmer must use the shortest possible information blocks, omitting sequence numbers, unnecessary words, and leading or trailing zeros where possible. Another trick is to use a sufficiently slow feed rate for the circumstances. The best solution is a rapid photoelectric tape reader instead of a mechanical one. Mechanical readers can read at rates of 60 to 100 characters per second. Photoelectric readers can read 300 to 500 characters per second.

7.8. STANDARD ADDRESS CODES FOR NUMERICAL CONTROL

- A   angular dimension about X axis
- B   angular dimension about Y axis
- C   angular dimension about Z axis
- F   feed function
- G   preparatory function
- H   unassigned (occasionally used for sequence number)
- I   center offset dimension in X in circular interpolation
- J   center offset dimension in Y in circular interpolation
- K   center offset dimension in Z in circular interpolation
- L   not used
- M   miscellaneous function
- N   sequence number
- O   not used, confusion with zero (occasionally used for sequence number)

S   spindle speed
T   tool number
U   secondary motion dimension parallel to X (i.e., a second tool turret or second machine head)
V   secondary motion dimension parallel to Y
W   secondary motion dimension parallel to Z
X   primary X motion dimension
Y   primary Y dimension
Z   primary Z dimension

7.9. PREPARATORY FUNCTIONS FOR CONTOURING MACHINES

G00   positioning control mode
G01   linear interpolation, straight line cutting
G02   circular interpolation, clockwise
G03   circular interpolation, counterclockwise
G04   dwell
G05   hold
G06, G07   unassigned. For special operations
G08   acceleration
G09   deceleration
G17   XY plane (see Fig. 59)
G18   ZX plane
G19   YZ plane
G33   thread cutting, constant pitch
G34   thread cutting, constantly increasing lead
G35   thread cutting, constantly decreasing lead
G40   cancel cutter compensation
G41   cutter compensation, cutter to left of surface (Fig. 60)
G42   cutter compensation, cutter to right of surface

7.10. CONTOURING MACHINES

Numerically controlled lathes are a special group of contouring machines, to be discussed in the following chapter. In this chapter the methods of contour programming will be discussed for the Gorton Tapemaster, the Kearney & Trecker Milwaukee-Matic Model II, and the Cintimatic family of mills and turret drills. Machines equipped with the Slo-Syn contouring controls are discussed in Chapter 9. Space does not permit a discussion of the interesting Compudyne Contoura, a contouring mill in use in many training schools and colleges.

There are minor differences in the methods used to program contours on these machines. The differences are of less importance than the similarities. All the above machines, when equipped for circular interpolation, require I, J, and, where used, K words for arc center offset, as explained in Sec. 7.3. All use incremental positioning commands when contouring, with the exception of the Cintimatics. Dimension words may be four digits, as for the Gorton Tapemaster, five digits, or six digits. Some differences in G (preparatory) and M (miscellaneous) functions will be noted.

When using incremental positioning, movements from the previous position in any axis must be coded plus or minus. To determine the sign of the dimension, the work table is assumed to be fixed in place and the toolhead is assumed to be numerically controlled. Actual conditions may well be the reverse; usually the table moves. Nevertheless, the sign convention assumes that the toolhead moves. The sign convention is illustrated in Fig. 61 and tabulated thus:

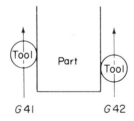

Figure 60

Axis	Tool movement	Table movement	Sign
X	to the right	to the left	+X
X	to the left	to the right	−X
Y	away from operator	toward the operator	+Y
Y	toward the operator	away from the operator	−Y
Z	upward	downward	+Z
Z	downward	upward	−Z

To illustrate, a movement of the tool 6.000″ to the right is coded X+06000.

In setting up a work piece for a contouring operation, a work piece zero reference point is required. This is the point at which the tool is located before operations are begun.

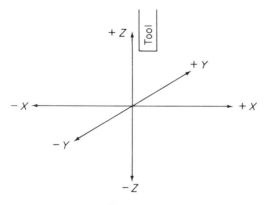

Figure 61

7.11. THE GORTON TAPEMASTER 2-30 CONTOURING MILL

This machine, illustrated in Fig. 62, is a positioning-contouring mill, in a two- or three-axis version, equipped with the Bunker-Ramo 3100 Control. Dimensions are incremented by using the sign convention explained in the previous Sec. 7.10. Acceleration and deceleration during movements are automatically controlled.

The tape format is word address. Trailing zeros may be omitted. The complete information block contains the following words:

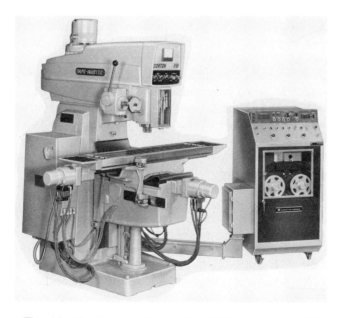

Fig. 62. The Gorton Tapemaster 2-30 contouring mill.

N000 G00 X±0000 Y±0000 Z±0000 I0000 J0000 F00 S00 M00 (EOB)

As usual, I and J words do not require a sign.

Incremental dimensions for the Gorton Tapemaster are interpreted by means of a G function. Dimension words, including any I and J in the information block, contain four digits. With a G01 preparatory function, the decimal point precedes the digits. With a G10 preparatory function, all dimensions in the block, including I, J, and K, are multiplied by 10, while a G60 function multiplies the dimension increments by 100. Thus,

G01   X+6   Y−0512   Z+375

means an X movement of the tool 0.6000 in. to the left, a Y movement of the tool of 0.0512 in. toward the operator, and a Z movement of 0.375 in. upward (or table downward). With a G10 function, these dimensions would be 6.000, 0.512, and 3.75, respectively. Therefore, the largest number that can be punched into the tape is 9999, and the maximum departure distance that can be programmed is 99.99, though this movement exceeds the capacity of the machine. Note that in the G60 mode the minimum increment is 0.01 in. G10 and G60 must be coded in every block where they are required; otherwise, control reverts to G01.

Coding of spindle speeds requires the use of a spindle speed table, which will not be reproduced here, and an M function: M42 commands a spindle start in the high-speed range and M43 a spindle start in the low-speed range. This method of coding spindle speed by an arbitrary series of numbers is not unusual.

Other M functions are M00—program stop for tool change or any other purpose, and M80 (instead of the standard M02) for end of program and tape rewind.

Normally when one is programming I and J arc center offsets, the I word precedes J. When programming the Gorton Tapemaster, J may be required to precede I in the information block. The order in which these two-dimension words are to be programmed is given in Fig. 63.

7.12. FEED RATE NUMBER

Feed rates along a tool path in contouring are coded by at least three methods:

1. Direct coding. In this method the actual feed rate in inches per minute is punched into the tape in the required number of digits. Thus F12 would command a feed rate of 12 or 120 or 1.2 ipm.
2. Feed rate number, FRN. This method is used with lathes, the Gorton Tapemaster, and many other machines.
3. Magic Three method. This is another method rather widely

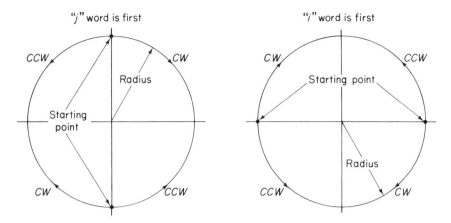

Figure 63

used. The Magic Three method is available as an option for the Cintimatic contouring machines to be discussed later in this chapter.

Feed rate number uses the following formula:

$$\text{FRN} = \frac{\text{feed rate in ipm}}{\text{length of cut}}$$

Frequently in this method, the feed rate in ipm is required to be multiplied by 10; that is,

$$\text{FRN} = 10\frac{F}{L}$$

In circular interpolation

$$\text{FRN} = \frac{F}{R} \text{ or } 10\frac{F}{R}$$

where R is the radius of the arc being contoured.

As an example, suppose that in linear interpolation a length of 8 in. is to be milled at 12 ipm. Then FRN = $\frac{12}{8}$ = 1.5, or for some machines FRN = $12 \times \frac{10}{8}$ = 15. This FRN of 1.5 or 15 must be coded in the required number of digits, usually 2 or 3.

In the Magic Three method of coding feed rate, the number must have three digits. The first digit is 3 greater than the number of digits to the left of the decimal point. The following two digits are the feed rate to two-digit accuracy.

For example, consider a feed rate of 1.0 ipm. There is one digit to the left of the decimal point. Code the first Magic 3 digit as 4 ($=1 + 3$). The following two digits are 10. A feed rate of 1 ipm is thus coded 410 in Magic Three.

Next consider a feed rate of 19.7 ipm. The first digit must be 5 ($=2 + 3$). The following digits are 20 (two-digit representation of 19.7). This feed rate is coded 520.

The Gorton Tapemaster uses the FRN method of coding feed rates, using the formula

$$\text{FRN} = \frac{\text{feed rate ipm}}{\text{length of cut}} \quad \text{or} \quad \frac{\text{feed rate}}{\text{arc radius}}$$

or in the G10 mode,

$$\text{FRN} = \frac{10 \times \text{feed rate ipm}}{\text{length of cut or radius}}$$

If a rapid traverse is required, then no feed rate is programmed.

7.13. A PROGRAMMING EXAMPLE FOR THE GORTON TAPEMASTER

The aluminum base of Fig. 64 is to be machined on a Gorton Tapemaster. The top surface is to be milled all round, and also a circle is to be milled in the middle of the part. The figure shows the zero reference point or parking position of the end mill.

The program for this operation is given in Fig. 65. The spindle speed is simply coded here as a fictitious number 11. Note also that the I and J dimensions must follow the order prescribed, and that these dimensions are incremented by 1 to ensure correct operation of the Tapemaster (J0000 is coded J0001). A feed rate of 14 ipm is selected, and used to calculate FRN formula. Note also that because the tool is returned to its parking position, the sum of all plus and minus departures in any axis sums to zero.

7.14. SLOPE AND ARC CUTS ON THE MODEL II MILWAUKEE-MATIC

The Model II of Fig. 30 uses absolute dimensioning when positioning and increments when contouring. The programming of departures of slopes and arcs follows standard procedures for contouring, except that six digits are used (four decimal places), including I, J, and K words. Trailing zeros are omitted, and I, J, and K words do not require plus or minus signs.

The plus and minus directions for contouring departures on the

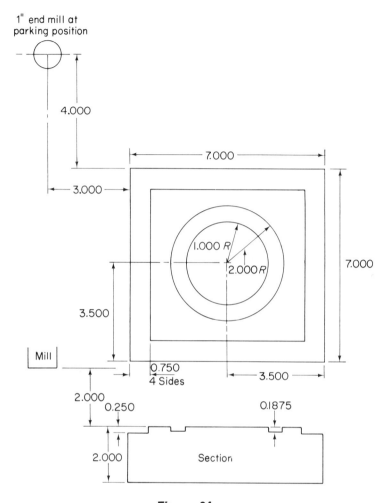

Figure 64

Milwaukee-Matic are not the standard ones of Fig. 61 but those suggested by Fig. 33. The +X, +Y, and +Z directions are those away from the zero position of each axis; −X, −Y, and −Z are departures toward the zero position. The plus sign may be omitted.

In the program sheet of Fig. 38, the last three columns, contouring feed rate and arc center offsets, are required only for a contouring machine. The method of determining the contouring feed rate will not be discussed here. It is a variant of the FRN method explained in Sec. 7.12. Since circular interpolation may be performed only in the XY, XZ, or YZ planes but not in all three, only two columns are required for arc center offsets, supplying either I and J, I and K, or J and K. As is usual in circular interpolation, only one quadrant of an arc may be programmed in any one information block.

PROGRAM SHEET

PART NUMBER AND NAME ALUMINUM BASE								GORTON TAPEMASTER DATE	
N	G	X	Y	Z	I or J	I or J	F	S	M
001									00
002								11	42
003	10	+3250	−3625	−2250			000		
004	10		−7125				020		
005	10	+6500					022		
006	10		+6500				022		
007	10	−7250					020		
008	10	−2500	+4250	+2250			000		
009	Tool change −								00
010	10	+6500	−6000	−1900			000	11	42
111	01			−2875			001		
012	10	−1500	−1500		J1501	I0001	100		
013	10	+1500	−1500		I1501	J0001	100		
014	10	+1500	+1500		J1501	I0001	100		
015	10	−1500	+1500		I1501	J0001	100		
016	01			+2875			000		
017	10	−6500	+6000	+1900			000		
018									00
019									80

Figure 65

A preparatory command 10, 11, or 12 is required for linear interpolation (see sequence no. 42 of Fig. 38), with six possible preparatory commands for circular interpolation (sequence nos. 59 and 60 of Fig. 38). The maximum radius or length of slope determines which preparatory command to use. The preparatory command 00 puts the machine in the positioning mode with absolute positioning, as in sequence nos. 1, 43, and 61.

Sequence nos. 41 and 58 show a preparatory function 44 and an X dimension of 01. Function 44 calls up contouring in the XY plane, with a dwell time to prepare the machine to receive contouring commands. The dwell time is the number of seconds shown in the X dimension, and must be a minimum of 1 second (X01). Function 45 calls up contouring in the XZ plane, and 46 in the YZ plane. Again a minimum of 1 second is programmed in the X column.

7.15. THE CINTIMATIC FAMILY OF NUMERICALLY CONTROLLED MACHINES

The Cintimatics as a family of n/c machines are available in two-, three-, or four-axis control, and in positioning and contouring systems. Most positioning mills, including the Milwaukee-Matics, use absolute positioning from a zero reference point for positioning and incremental programming for contouring operations. The Cintimatics are somewhat unusual in that absolute positioning is employed also when contouring. A considerable number of canned cycles (G functions) are used by these machines, and feed rate may be coded by various methods, including Magic Three.

The horizontal mill, vertical mill, and turret drill, Fig. 18, will be discussed in the following pages. Since the intent is to discuss these machines as a group, insofar as that is possible, complete programming details for any specific machine of the group will not be supplied. Such details as table indexing, deep hole drilling, and the several bore functions G85 to G89 are omitted.

The simple two-axis vertical spindle positioning mill was discussed in Chapter 3. We consider next the same basic mill, with either horizontal or vertical spindle, and Acramatic 225-W machine control unit, with or without contouring control.

These mills use the following M codes:

M00 mandatory stop
M01 optional stop
M02 end of program
M06 tool change
M26 full spindle retraction from gage height

With the exception of M26, these functions have been explained in Chapter 3.

The function M26 calls for spindle retraction from gage height to fully retracted position. Gage height is the terminal point for the rapid advance and the initial feed point, as explained in Chapter 3. In the case of the horizontal and the vertical mill, but not the turret drill, the depth of rapid advance and of final feed position is set up on cams. The standard G functions such as G81 (drill cycle) terminate with the tool retracted to the gage height. If a series of holes is to be drilled, the drill retracts to gage height each time after drilling. At the end of the program, an M26 fully withdraws the quill. The quill may also be withdrawn by programming M06.

Many of the preparatory G functions of these machines have been discussed in Chapter 3.

G78 mill stop. This function is not available for the turret drill.
G79 mill cycle. A G79 must be programmed for linear interpolation.
G80 cancel cycle
G81 drill cycle
G84 tap cycle
G85 bore cycle

126   *Chap. 7 / Basic Principles of Contouring*

Other special bore cycles are available, but are not discussed here.

The configuration of machine axes for each of these machines is shown in Fig. 66. Only two-axis mills are discussed, and therefore no Z axis is shown for these machines. The zero point may be shifted in X or Y or both to any location, and hence there may be negative X and Y coordinates. The sign of X, Y, I, and J, need be recorded only when it changes.

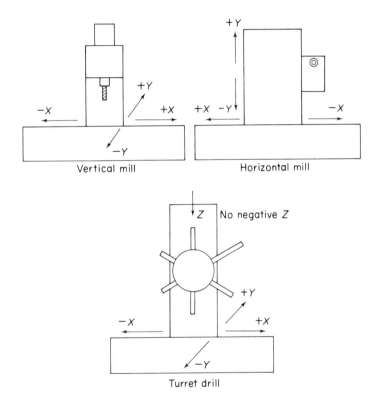

Figure 66

7.16. PROGRAMMING THE CINTIMATICS

Linear interpolation on these machines is restricted to the XY plane. Suppose that the tool path of Fig. 67 must be programmed.

A G79 preparatory function must be used if X and Y feed rates are to be coordinated so as to cut a straight line at any desired angle.

If this machining sequence is executed on either milling machine,

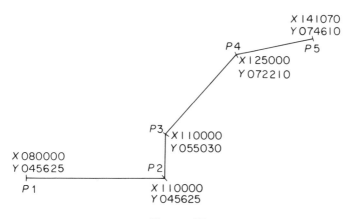

Figure 67

the program will be that of Fig. 68. The dimensioning is of course absolute, from the zero point. Seven digits are used for dimensioning and four digits for feed rate (one decimal place). Leading zeros may be omitted. Since the format is word address, the tape coding would be the following:

```
H023   G79   X+80000    Y+45625   F100   W01   (EOB)
N024         X  110000
N025                    Y55030
N026         X125100    Y72210
N027         X141070    Y74610                 M26   (EOB)
```

F is the letter address for feed rate and W for cam selection, with cam numbers W00 to W09. W00 selects manual cam operation, while W01 to W09 call up cams 1 to 9. As described in Chapter 3, each cam has a setting for quill rapid advance limit and a second setting for quill feed limit. It will be concluded then that these Cintimatic mills with broader capabilities are programmed basically like the simpler mill of Chapter 3.

Figure 69 shows the program manuscript for the same operation as performed on the turret drill with Acramatic 335-D machine control unit. A few of the programming details for the turret drill require explanation. In common with the mills, leading zeros may be omitted. Letter addresses are

Z feed position	— Z (7 digits, 4 decimal places)
Z rapid position	— R (7 digits, 4 decimal places)
Feed rate	— F (4 digits, 1 decimal place)
Spindle speed	— S (3 digits for Magic Three coding)
Tool number	— T (2 digits, leading zero may be omitted)
Arc center offsets	— I, J (7 digits, 4 decimal places)

PART									CINTIMATIC MILL	
SEQ NO.	G	X	Y	I	J	F	CAM (W)	M	REMARKS	
023	79	+0080000	+0045625			0100	01		P1	
024		0110000							P2	
025			0055030						P3	
026		0125100	0072210						P4	
027		0141070	0074610					26	P5	

Figure 68

PART											CINTIMATIC TURRET DRILL	
SEQ NO.	PREP FCN	X	Y	Z	I	J	F	R	S	T	M	
023	80	+80000	+45625	00			100	3000		1		P1
024	79	110000										P2
025			55030									P3
026		125100	72210									P4
027		141070	74610									P5
028	80										26	P6

Figure 69

The miscellaneous functions 00, 01, 02, 06, and 26, discussed for the milling machines, apply also to the turret drill. In addition the following standard functions, discussed for the Milwaukee-Matics, are also used:

    M02   spindle on clockwise
    M04   spindle on counterclockwise
    M05   spindle off
    M07   mist coolant on
    M08   flood coolant on
    M09   coolant off.

7.17. Z AXIS MOVEMENTS FOR THE TURRET DRILL

The manuscript sheet for the turret drill contains two Z dimensions: Z feed (Z) and Z rapid (R). The R dimension is the terminal point of the rapid advance of the tool, measured from an R000 0000 position defined by the setup procedure shown in Fig. 70. Since there are no negative Z coordinates, R and Z do not require a plus sign when programmed. Now suppose the drill is to drill 1 inch into the work piece, as in Fig. 71. This operation is programmed

Z	R
10000	00

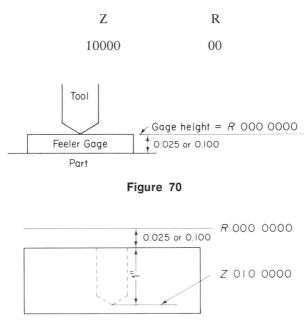

Figure 70

Figure 71

Here the Z dimension appears to be incorrect. If the R dimension—the top of the feeler gage—is zero, then feeding 1 in. beyond that point will give too shallow a hole by the thickness of the feeler gage. This programming, however, is correct for the Cintimatic turret drill, because the controller automatically supplies the additional feed equivalent to the gage height, and the hole is actually drilled to correct depth.

For a multilevel part, see Fig. 72.

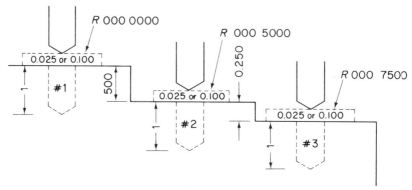

Figure 72

Now the setting of a gage height can apply to only one tool of specific length. Other operations will have to be performed on the part surface by tools which differ in length, and the R setting would not apply to tools of varying length. During setup, a tool compensation procedure is used such that all tools perform as if they were the same length. This is done by thumbwheel switches for each tool. When a specific tool number is called up, this tool number calls up also the required tool compensation.

In the program of Fig. 69, the tool is to advance rapidly to full cutting depth instead of to gage height, since the R position is 0.3 in. below the Z position (R is usually above Z). In this case only, the gage thickness must be added to the R value, so that

$$R = 0.3 = \text{gage thickness} + \text{depth of cut}$$

For all other operations, the gage thickness is ignored in the programming. Unlike the G80 function of Chapter 3, G80 here positions the tool rapidly at the XY position then rapidly to gage height. There is no Z feed rate motion in the G80 cycle, for the turret drill or for the Cintimatic mill of Chapter 3.

7.18. UNIT OPERATIONS WITH THE TURRET DRILL

To make clear the operation of the Cintimatic turret drill, a few typical operations are explained. These practices would apply to many other turret drills.

1. MULTILEVEL DRILLING. In Fig. 72 the work piece has three levels, all of which are to be drilled to a depth of 1 in. The R and Z values for the drilling operation are these:

Hole	R	Z
#1	000 0000	001 0000
#2	000 5000	001 0000
#3	000 7500	001 0000

To repeat, the gage height required to reach the total feed distance for full depth of hole is generated in the control and is not programmed. A gage height of either 0.025 or 0.100 must be selected, and both programmer and machine operator must agree on the selection of gage height if the proper depth of hole is to result.

2. **BASIC MILL CYCLE G79.** See Fig. 73. In sequence no. H010, G79 moves the work piece to position 1 at the programmed feed rate of 8 ipm, as given in the program of Fig. 74. At position 1 the tool moves rapidly to the R00 plane, then feeds to depth at the programmed feed rate.

In sequence N011 the G79 cycle is maintained to move the work at 8 ipm to position 2, with a depth of cut of 0.15 in. At sequence 012, the

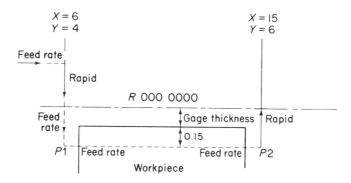

Figure 73

PART								CINTIMATIC TURRET DRILL			
SEQ NO.	PREP FCN	X	Y	Z	I	J	F	R	S	T	M
010	79	+60000	+40000	1500			80	00	600	1	03 P1
011		0150000	60000								P2
012	80										

Figure 74

G80 command retracts the tool to gage height at R00. If full retraction is desired, then either M06 or M26 would be programmed in this last information block.

3. CANCEL CYCLE G80. When this cycle is programmed, the tool moves at rapid traverse to the XY position in the information block, then rapidly advances to gage height. This cycle is slightly different from the G80 of Chapter 3. See Fig. 75.

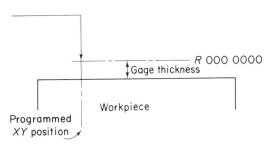

Figure 75

4. DRILL CYCLE G81. See Fig. 76, with the associated program of Fig. 77. In sequence 010, G81 calls for the drill to move rapidly to the XY position of the first hole. At this position the drill advances rapidly to gage height, which is the R00 plane. Next the drill feeds in at the programmed feed rate. At the Z depth of 1.15 in. the drill retracts rapidly to gage height R00. At sequence 011, the drill moves rapidly to the second hole location, feeds in to 1.00 in. and rapidly retracts to R00 above the part. The M26 command causes the drill to retract fully.

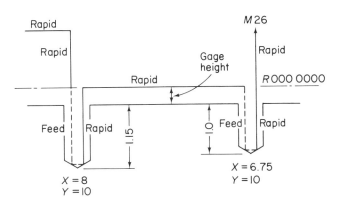

Figure 76

PART								CINTIMATIC TURRET DRILL		
SEQ NO.	PREP FCN	X	Y	Z	I	J	F	R	S	T	M
010	81	+80000	+100000	11500			70	00	500	2	03
011		+67500		10000							26

Figure 77

5. CHANGING LEVELS IN THE WORK PIECE.

In Fig. 78 the first hole must be drilled at a higher elevation than the second hole. In the program for this operation, Fig. 79, note the difference in gage elevation, the second R plane being 1 in. lower than the first. The tool will retract rapidly to the gage height at each hole.

If the second or lower hole is to be drilled first and then the first hole, the two drilling operations cannot be performed in two blocks of information both with G81, since the second block will call for a rapid movement to the next drilling position while the tool is at the lower R elevation. A tool-workpiece collision will result. The programming for such a case is given in Fig. 80. Sequence 021 retracts the tool to the new R value given in this block, which must not contain any new XY coordinates.

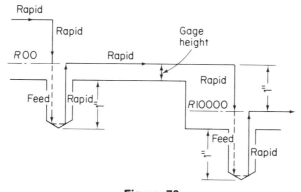

Figure 78

134 *Chap. 7 / Basic Principles of Contouring*

PART											CINTIMATIC TURRET DRILL	
SEQ NO.	PREP FCN	X	Y	Z	I	J	F	R	S	T	M	
020	81	+80000	+110000	10000			80	00	400	3	03	P1
021		135000	100000					10000				P2

Figure 79

PART											CINTIMATIC TURRET DRILL	
SEQ NO.	PREP FCN	X	Y	Z	I	J	F	R	S	T	M	
020	81	+135000	+100000	10000			80	10000	400	3	03	P2
021	80							00				P2
022	81	80000	110000									P1

Figure 80

8

THE NUMERICALLY CONTROLLED LATHE

8.1. CONVENTIONAL PRACTICE FOR N/C LATHES

Numerically controlled lathes are available in several types: chucking and center-turning lathes, turret lathes, and in two, three, or four axes. Axis nomenclature is given in Fig. 81.

A Warner & Swasey two-axis (Z longitudinal and X cross) chucking lathe equipped with a six-station tool turret is illustrated in Fig. 1. The placing of the ways in a vertical plane is a common practice in n/c lathe construction. This configuration has the obvious advantage that chips are kept from falling on the ways. Another advantage is in reduced force against the ways and therefore reduced wear; the top plane of the ways carries the dead weight of the tooling, and the side plane receives the cutting force. With horizontal ways these two forces are additive.

An n/c lathe costs more than a comparable engine lathe. The cost of a lathe is recoverable only as fast as the lathe makes chips. Therefore, the recovery of the first cost of an n/c lathe within a reasonable time is possible only if it has adequate chip-making capacity. One motor horsepower can produce approximately a cubic inch of steel chips per minute, and a smaller installed horsepower than 15 hp would be unusual in an n/c lathe.

Numerically controlled lathes use incremental dimensioning. In incremental positioning the origin of coordinates moves with the tool from position to position. Plus and minus signs indicate the direction of in-

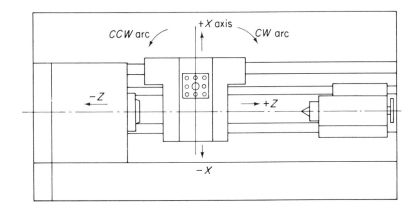

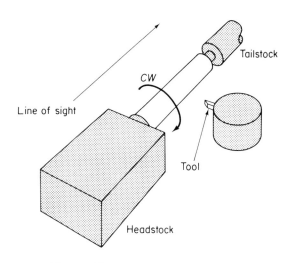

Figure 81

cremental cutter movement. As Fig. 81 shows, the following conventions hold:

+Z (or +W)—movement of tool carriage away from the headstock

−Z (or −W)—movement of tool carriage toward the headstock

+X (or +U)—movement of cross slide toward the operator (up, in the case of the Warner & Swasey and Monarch turN/Center)

−X (or −U)—movement of cross slide away from the operator (down, in the case of the vertical-plane Warner & Swasey and Monarch turN/Center)

As usual, the plus sign may be omitted.

The convention for clockwise (CW) and counterclockwise (CCW) spindle rotation is the usual one, that is, as viewed from the headstock toward the tailstock.

Dimension words for most lathes have five digits, with the decimal point understood to follow the *first* digit. Thus X06250 is 0.625 in., not, as for mills and drills, 6.250 in. Trailing zeros may be omitted, but leading zeros must be included.

For most lathes, the maximum distance that can be programmed in any block of information is 9.9999 in. If a longer distance must be cut, the total length is programmed in two equal distances in two successive information blocks. Thus, if the cut must be 11.5 in. long, it is programmed as two cuts, each 5.75 in. Two cuts of unequal length may be programmed, but this is slightly inconvenient for programming the feed rate number. Feed rate number is explained later.

Like most other n/c machines, n/c lathes may be operated manually, as they must be when one is setting up for a new part. Just as with a positioning mill, setting up a lathe job begins by referencing the cutter tip and the work piece to each other in an absolute dimensioning system. Manually adjustable tool offset dials are used to bring the tool point accurately to location, just as zero offset dials are available for mills. When a throwaway carbide tip is replaced with a new one, an error of a few thousandths of an inch in tool offset is introduced, and this error is removed by the tool offset dials.

The following word address format is common to virtually all n/c lathes:

Word	Address	Digits
Sequence number	N	3
Preparatory function	G	2
Cross dimension	X	5
Longitudinal dimension	Z	5
Arc center offset, X	I	5
Arc center offset, Z	K	5
Feed rate	F	4
Spindle speed	S	3
Tool number	T	2
Miscellaneous function	M	2

In general, a word read into storage remains in storage until erased by a new word read into the same address, with the usual exception of M00, M01, and M02.

8.2. DIMENSIONING THE TOOL PATH

N/c lathes are contouring machines. A complicating factor is the substitution of programming procedures for machining skills, and surely no one better conceives how much skill there is in the machinist's trade than the person who has just completed his first n/c lathe program.

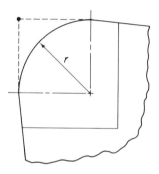

Figure 82

Except for such cutters as single-point threading tools, lathe tools have a nose radius. This nose radius is the cause of somewhat irksome but necessary trigonometric computations for tool path. Consider the tool contour of Fig. 82, where the cutter has a radius r. If it were a pointed tool, it would have the contour of the dashed lines, and the path of the tool point would be the actual shape of the part to be made in the lathe. Now consult Fig. 83, which shows an external radius to be turned. A sharp-pointed tool could produce the actual curve as required. The surface finish, of course, would be unacceptable, and the tool life equally unacceptable. Therefore, the tool must be radiused. If, however, the radiused tool is programmed as a sharp-pointed tool, it cuts the curve shown as a dashed line, cutting first on one side of the tool, then at the end of the cut on the trailing side of the tool. As a result, there is a deviation between the desired curve and the actual curve produced, the maximum deviation occurring at 45 deg. The maximum error (at 45 deg) produced by various tool radii is indicated in the graph of Fig. 84. As a rule of thumb, this error maximum is half the tool radius; thus, for the common tool radius of $\frac{1}{32}$ or 0.031, a maximum error of about 0.015 in. results. Sometimes this deviation can be ignored, but for most lathe programming, tool radius must be allowed for.

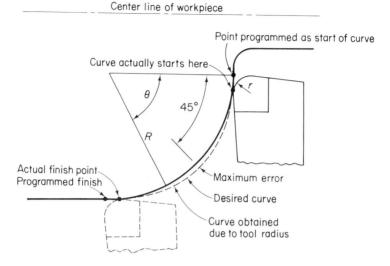

Figure 83

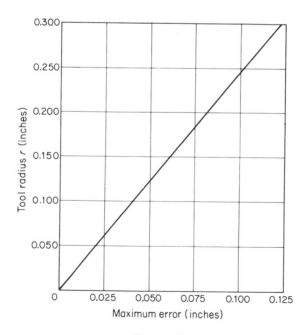

Figure 84

Thus in contouring a part, except for straight cuts, different segments of the tool nose are in contact with the work piece at various times, resulting in dimensional errors. This difficulty is surmounted by programming the movement of the center of the tool radius instead of the movement of the tool nose. This is similar to the case for end milling, where again the center of the cutter is programmed. When one is contouring an arc such as a fillet, the arc followed by the cutter is the part radius R plus or minus the tool radius r. For a fillet it will be $R - r$; for the external radius of Fig. 83 it will be $R + r$.

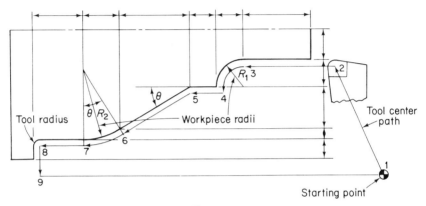

Figure 85

A typical case of the relationship of work piece contour and tool path is illustrated in Fig. 85. Anyone learning to program an n/c lathe will require such a layout sketch so that his thinking will be correctly oriented to the actual tool path. Some prefer to draw tool circles at the programming points 1, 2, 3, etc.

For cuts parallel to X and Z axes, the relationships are shown in Fig. 86, which is a case of a stepped work piece. The dimensions for X

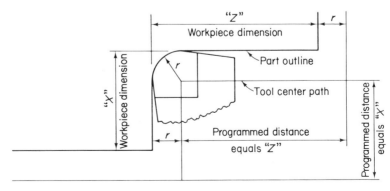

Figure 86

and Z in such a case can be taken from the work piece, though the initial and terminal points of tool movement are displaced from the steps in the shaft by the tool radius.

8.3. DIMENSIONING TOOL DISPLACEMENTS FOR TAPER CUTS

To resolve the problem of nose radius for taper cuts, consult Fig. 87, which compares a sharp and a radiused tool. The figure shows that

$$D = \text{horizontal error} = r - r \tan \frac{\theta}{2}$$

Similarly,

$$d = \text{vertical error} = r - r \tan \frac{\phi}{2}$$

For convenience, $r \tan \theta/2$ will be designated K_z, and $r \tan \phi/2$ K_x.

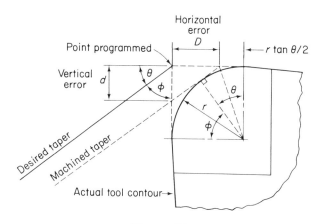

Figure 87

All possible cases of intersections of tapers with cuts parallel to either machine axis are illustrated in Fig. 88(a), (b), (c), (d). The tool displacements to be programmed are given with each figure, and these should be reasonably self-evident.

Chamfer cuts can be handled with the data of Fig. 89, which is taken from the excellent programming manual for the turN/Center 75 lathe made by Monarch Machine Tool Company, Sidney, Ohio, 45365.

Another possibility in taper calculations should be mentioned: the intersection of two tapers on a shaft. This requires slightly more complex

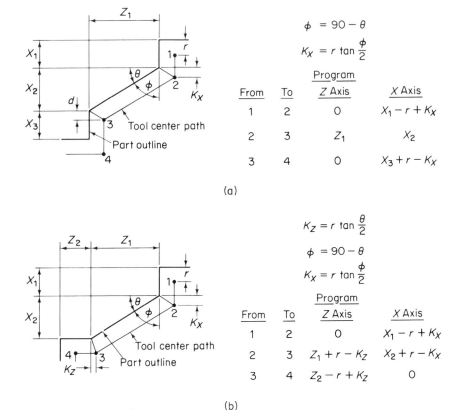

Figure 88

trigonometric calculations. This problem will not be discussed here, since it is not commonly encountered.

For tapers tangent to a circular arc, consult Fig. 90, which includes all possible configurations. In this figure, K_z means the same as before, $r \tan \theta/2$. Note that K_z is required at the end of the taper that joins a straight Z cut, and $r \sin \theta$ and $r \cos \theta$ where the taper meets a tangent arc.

This initial operation of determining tool center positions is the most cumbersome part of lathe programming.

8.4. THE MONARCH TURN/CENTER 75 TWO-AXIS LATHE

Because it is an easy machine to program, this lathe, manufactured by Monarch Machine Tool Company, Sidney, Ohio, will be used for explaining the procedures of lathe programming. A general view of the machine is given in Fig. 91, with its General Electric machine control unit.

The Monarch Turn/Center 75 Two-Axis Lathe

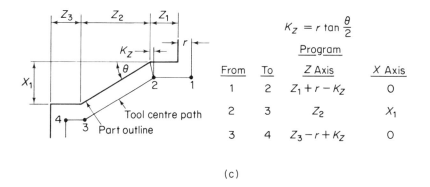

(c)

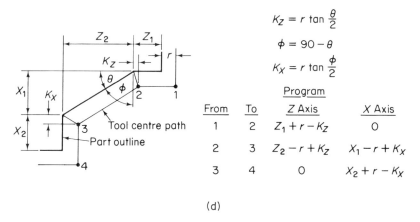

(d)

Figure 88 (cont.)

A four-station tool turret is supplied with this lathe. The following are other specifications for the machine:

Swing over bed, 20 in.
Distance between centers, 54 in.
Hole through spindle, $3\frac{1}{16}$ in.
Spindle speed range, 40–2000 rpm
Lathe tools $1 \times 1\frac{1}{4}$ in.
Minimum n/c increment, 0.0001 in.
Feed range, 0–100 ipm
Rapid traverse, 200 ipm
Drive motor, 20 hp
Weight, 12,750 lb

144 *Chap. 8 / The Numerically Controlled Lathe*

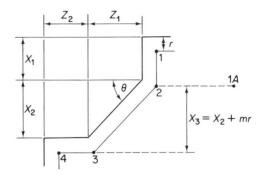

Program

From	To	Z Axis	X Axis
1	2	0	$X_1 - mr$
2	3	$Z_1 + nr$	$X_2 + mr$
3	4	$Z_2 - nr$	0

The multipliers m and n may be calculated from:

$$m = 1 - \tan\frac{90-\theta}{2}$$

$$n = 1 - \tan\frac{\theta}{2}$$

Tool Radius r	$\theta = 30°$ mr	$\theta = 30°$ nr	$\theta = 45°$ mr	$\theta = 45°$ nr	$\theta = 60°$ mr	$\theta = 60°$ nr
1/64	0.0066	0.0114	0.0092	0.0092	0.0114	0.0066
1/32	0.0132	0.0229	0.0183	0.0183	0.0229	0.0132
3/64	0.0198	0.0343	0.0275	0.0275	0.0343	0.0198
1/16	0.0264	0.0458	0.0366	0.0366	0.0458	0.0264

Figure 89

All n/c lathes use a programming increment of 0.0001 in.

The General Electric 7542 contouring control unit does not employ buffer storage but uses a high-speed photoelectric tape reader.

For the information block of the Monarch lathe, refer to the standard lathe information block of Sec. 8.1. This standard block corresponds to the tape information supplied to the Monarch 75, except for the following variations.

1. The Monarch 75 uses six-digit words for X, Z, I, and K, with the decimal point following the second digit. Thus Z241250 or Z24125 both are read 24.125 in.
2. The feed rate word F has five digits in the case of the Monarch

The Monarch Turn/Center 75 Two-Axis Lathe

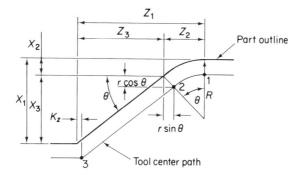

From	To	Z Axis	K Word	X Axis	I Word
1	2	$Z_2 - r\sin\theta$	0	$X_2 + r\cos\theta - r$	$R - r$
2	3	$Z_3 - K_z + r\sin\theta$	0	$X_3 + r - r\cos\theta$	0

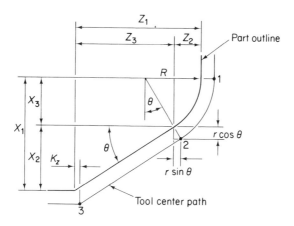

From	To	Z Axis	K Word	X Axis	I Word
1	2	$Z_2 + r - r\sin\theta$	$R + r$	$X_3 + r\cos\theta$	0
2	3	$Z_3 - K_z + r\sin\theta$	0	$X_2 + r - r\cos\theta$	0

Figure 90

machine, including two decimal places. Thus F20000 or F2 both mean 200.

3. The spindle speed is coded in two digits for the Monarch lathe.

This machine can read the following five G or preparatory commands, which are standard for all lathes. More complex lathes may use more than these.

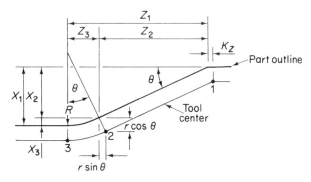

From	To	Z Axis	K Word	X Axis	I Word
1	2	$Z_2 + K_z - r\sin\theta$	0	$X_2 - r + r\cos\theta$	0
2	3	$Z_3 + r\sin\theta$	$Z_3 + r\sin\theta$	$X_3 + r - r\cos\theta$	$R - X_3 + r\cos\theta$

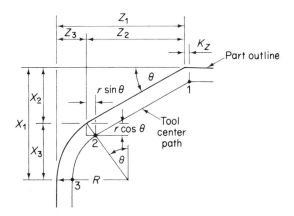

From	To	Z Axis	K Word	X Axis	I Word
1	2	$Z_2 + K_z - r\sin\theta$	0	$X_2 - r + r\cos\theta$	0
2	3	$Z_3 + r\sin\theta - r$	$R - Z_3 - r\sin\theta$	$X_3 - r\cos\theta$	$X_3 - r\cos\theta$

Figure 90 (cont.)

G01—linear interpolation. To be coded when a straight cut or a taper is programmed.

G02—circular interpolation, when the lathe cutter is to cut a clockwise arc.

G03—circular interpolation, when the lathe cutter is to cut a coun-

Fig. 91. The Monarch turN/Center numerically controlled lathe.

terclockwise arc. With the circular interpolation facility, only basic information about the arc is programmed, and the machine control unit calculates the tool positions to generate the arc.

G04—coded when a dwell is programmed. The time of the dwell in seconds is coded as an X dimension, though many other lathes code the dwell as a Z dimension. Thus G04X02 means a 2-second dwell. But G01X02 means $X = 2''$.

G33—threading operation. Only threads of constant pitch can be cut on this lathe. More complex lathes can cut threads of increasing or decreasing pitch.

All these G functions are held in storage until cancelled by a different G function.

The G02 and G03 commands refer to CW and CCW arc cuts. Refer to Fig. 81 for the sense of CW and CCW. For other n/c lathes equipped with ways in the more familiar horizontal plane, these directions as shown in the figure will be reversed because of n/c axis conventions, and an apparently CW arc must be programmed as CCW.

The following standard M or miscellaneous functions are programmed for this lathe:

M00 program stop
M01 optional stop, at the option of the operator, who must push a button before the M01 code is read from the tape
M02 end of program

The above three codes are not held in memory.

M03 spindle start CW
M04 spindle start CCW
M05 spindle stop
M08 coolant on (program M08 before the tool begins to cut. Putting cold coolant on a hot carbide is bad practice.)
M09 coolant off
M30 end of tape

FEED RATE NUMBER. The Monarch turN/Center 75 uses the same method of coding feed rate numbers as is used with most n/c lathes. This method is the feed rate formula

$$FRN = 10F/L$$

where F is the desired feed rate in ipm along the tool path and L is the length of the tool path, either the X or Z increment or the hypotenuse of a cut moving in both X and Z. A FRN greater than 500 cannot be programmed on many lathes.

As an example of FRN, suppose that the desired feed rate was 6 ipm and the cut was a taper, 12 in. in Z and 5 in. in X. The length of cut along the taper is then 13 in.

Then

$$FRN = \frac{10 \times 6}{13} = 46.154$$

The Monarch 75 requires that this be programmed as 04615, five digits, two of which are decimal places.

In the case of a circular arc, the FRN formula substitutes the radius of the arc for L; thus,

$$FRN = \frac{10F}{R}$$

There are a number of restrictions and manipulations that are applied to FRN's. These can be obtained from programming manuals for specific n/c lathes, but will not be discussed here because of the additional detail they would add to this chapter. Many lathes are restricted to feed rate numbers of 500 or less. In this chapter this restriction will be imposed on the Monarch turN/Center 75, although a more complex restriction rule applies to this machine.

Spindle speeds for n/c lathes use a number code, which is obtained from a table of speeds supplied in the programming manual for the machine. Here again, the coding of speeds for the Monarch turN/Center 75 is typical. The following is the table of spindle speeds for it.

turN/Center 75

Spindle Speeds

LOW		MED		HIGH	
Code	RPM	Code	RPM	Code	RPM
S00	40	S20	92	S40	222
S01	45	S21	105	S41	253
S02	51	S22	119	S42	287
S03	58	S23	135	S43	326
S04	66	S24	153	S44	371
S05	75	S25	174	S45	322
S06	86	S26	198	S46	479
S07	97	S27	225	S47	545
S08	111	S28	252	S48	619
S09	126	S29	291	S49	704
S10	143	S30	331	S50	800
S11	158	S31	366	S51	886
S12	175	S32	406	S52	981
S13	194	S33	449	S53	1086
S14	215	S34	497	S54	1202
S15	238	S35	550	**S55**	**1331**
S16	263	S36	609	S56	1474
S17	291	S37	675	S57	1632
S18	322	S38	747	S58	1806
S19	357	S39	827	S59	2000

There are three speed ranges, low, medium, and high, and to shift from one range to another requires a gear shift. During gear shifting the spindle is stopped. Therefore, if a gear shift from one range to another range is programmed, the tool must be retracted (so that the work piece is not marked) and a 2-second dwell is programmed as X02. However, a speed change within any speed range may be made while the tool is cutting.

150   Chap. 8 / *The Numerically Controlled Lathe*

Thus to change from 92 rpm (medium range) to 105 rpm does not require a gear shift and dwell, but a change from 252 in the medium range to 322 in the high range requires a dwell.

The tape coding for speeds is read from the table. For 609 rpm, punch S36; for 215 rpm, punch S14.

The tool word T calls up the tool number in the first digit and one of the four tool offsets in the second digit. Thus T13 means tool no. 1 and tool offset no. 3.

All X, Z, I, and K dimensions are reset to zero at the end of each information block. The I and K arc center offset words will be discussed presently; this would seem to be the place to bring all these programming details together in an example of n/c lathe programming.

8.5. EXAMPLE 1: STRAIGHT TURNING

The work piece to be turned is illustrated in Fig. 92. To simplify the example, the part is assumed to be rough-turned. The shaft is held between chuck and tailstock center. It is not parted off in the lathe, and no chamfers are programmed.

The shafting material is 0.4 per cent carbon steel. The steel mill that supplies the steel recommends a cutting speed of 585 fpm using disposable carbides. There are two shaft diameters to be turned, $1\frac{1}{4}$ and $1\frac{3}{4}$.

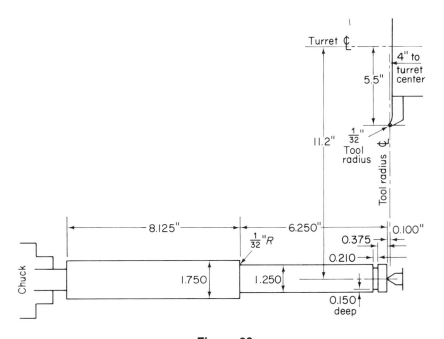

Figure 92

The periphery of a 1¼-in. shaft measures about 4 in. or ⅓ foot, so to obtain a cutting speed of 585 fpm requires 1755 rpm ideally. For the 1¾-in. diameter an rpm of 1270 is desired. Both these rpm's lie in the high-speed range of the Monarch 75. Select speeds of 1632 and 1202 rpm, coded respectively S57 and S54. Since both speeds lie within the same high range, no gear change is required, and the speed change can be made while the tool is turning the shaft.

There is also a plunge cut for an O-ring groove at the end of the shaft. The cutting speed selected for this operation is 300 fpm, or about 900 rpm. Speed S51 could be selected. But in order to show the programming for a gear shift, speed S39 (827 rpm) will be selected because it is in the medium-speed range.

The following cutting sequence will be adopted:

1. Turn 1¼ diameter, using tool T11.
2. Face 1¾ shoulder, using tool T11.
3. Turn 1¾ diameter, using tool T11.
4. Retract tool turret, index tool, gear shift.
5. Plunge cut groove, using tool T34.
6. Retract tool turret.

The location of the plunge tool in the turret is not shown in the figure. Assume that it projects 5.5 in. from the turret center line to the cutting face of the tool, and that if indexed into position, its left-hand corner would occupy the position of the center of the nose radius of tool no. 1.

The tool turret is shown in the figure in the fully retracted position in X. The turret may be set up in any "home" position in the Z direction. In this example it is set up 0.100 in. to the right of the work piece, as measured from the center of the tool radius.

8.6. THE PROGRAM

The n/c program for the operation is given in Fig. 93.

BLOCK 001

A 2-second dwell, G04 X02, is programmed to index T1 into position and to engage and accelerate the spindle.

BLOCK 002

The tool makes a rapid advance to cutting position at the required work piece diameter, moving only in −X. The X departure is given by

$$11.2 - 5.5 - 0.625 - 0.031 = 5.044 \text{ in., coded X}-05044$$

where 11.2 = distance from turret center to work piece center
5.5 = tool offset in turret
0.625 = radius of finished shaft
0.031 = tool nose radius

Since the movement in X is a rapid advance, the maximum velocity of 200 ipm is used.

$$\text{FRN} = \frac{10 \times 200}{5.044} = 396.51, \text{ coded F39651}$$

The coolant is turned on with M08.

BLOCK 003

The G01 function for straight-line movements still holds and is not repeated. The tool moves only in $-Z$ a distance equal to

$$6.250 + 0.100 - 0.031 = 6.319 \text{ in.}$$

The feed is 0.008 in./revolution, or 13 ipm.

$$\text{FRN} = \frac{10 \times 13}{6.319} = 20.57$$

BLOCK 004

Next the shoulder must be faced, going 0.250 in. in $+X$, with spindle speed dropping to S54. Using an estimated feed rate of 10 ipm for the lower spindle speed, we have

$$\text{FRN} = \frac{10 \times 10}{0.25} = 400$$

BLOCK 005

The $-Z$ departure is

$$8.125 + 0.031 + 0.100 = 8.256 \text{ in.}$$

The 0.100-in. distance is sufficient for the tool to pass clear of the work piece.

$$\text{FRN} = \frac{10 \times 10}{8.256} = 12.11$$

BLOCK 006

In this block the tool will be retracted from the work piece in order to position it for the plunge cut and also so that a gear shift can be pro-

MONARCH PATHFINDER LATHE NUMERICAL CONTROL PROGRAM SHEET

OPERATION: _____ DATE: _____ SE NO.: _____ HOME POSITION: Long. _____ Trans. _____
PLANNER: _____ PART NO.: _____ MATERIAL: 1040 STOCK SIZE: ROUGH MACHINED
CUSTOMER: _____ PART NAME: REVERSE SHAFT HARDNESS: _____ MACHINE: TURN/CENTER 75
TAPE NO.: 0001

ABSOLUTE DIMENSION FROM START		SEQ. NO.	PREP. FUNC.	INCREMENTAL DISTANCE			DISTANCE TO ARC CENTER		FEED FUNCTION	SPIN. SPEED	TOOL NO.	MISC. FUNC.	OPERATION, TOOL HOLDER, AND TOOL HOLDER LOCATION	SFPM	RPM	DIA.	FEED		DEPTH OF CUT
x axis	z axis			Transverse	Longitudinal	z±	Parallel to X	Parallel to Z									IPR	IPM	
n			g	x±			i	k	f	s	t	m							
0	0	001	04	02								03	INDEX TOOL, SPINDLE CW			1¼			
		002	01	-05044	-06319				39651	57	11	08	RAPID TO WORK						
-5.044	-6.319	003							02057				TURN 1¼ DIA	585	1632		008	13	
		004		0025					400	54			FACE SHOULDER				008	10	
-4.794	-14.575	005			-08256				01211				TURN 1¾ DIA.	585	1202		008	10	
		006		03	1.389				14075				RETRACT FOR PLUNGE						
-1.794	-0.685	007	04	02									SPEED CHANGE						
-4.975		008	01	-03181					5	39	34		RAPID TO WORK						
-5.225		009		-0025					064				GROOVE		827		002	1.6	
0		010		05225					38469			09	RETRACT IN X						
	0	011			00685				5				POSITION IN Z						
		012	04	02						57	11		SPEED CHANGE						
		013										05							
		014										02							

Fig. 93. Program for reverse shaft of Figure 92.

grammed in the following block. The retraction in +X is decided as 3 in., and in +Z to the position that places the plunge cutter opposite the groove it is to cut. The Z departure is

$$0.100 + 8.125 + 6.250 - 0.210 - 0.375 = 13.890 \text{ in.}$$

$$\text{FRN} = \frac{10 \times 200}{14.21} = 140.75$$

The hypotenuse of both X and Z movements is 14.21.

BLOCK 007
Tool change and spindle change during a 2-second dwell.

BLOCK 008
The tool cutting edge is now positioned vertically above the groove to be cut, at a distance of 3.906 in. above the spindle axis. The tool is rapidly advanced to 0.100 in. from the work surface, that is, to a distance of 0.275 from the spindle axis.

$$\text{FRN} = \frac{10 \times 200}{3.181}$$

Use 500.

BLOCK 009
This is the plunge cut. Using a feed of 0.002 in. per revolution gives 1.6 ipm. The depth of cut or −X departure is 0.250 in.

$$\text{FRN} = \frac{10 \times 1.6}{0.25} = 64$$

BLOCK 010
The tool will be retracted to the X position from which this lathe program began. The X departure is 5.225 in.

$$\text{FRN} = \frac{10 \times 200}{5.225} = 384.69$$

BLOCK 011
The tool turret is shifted in +Z to its initial position from which this lathe program began.

BLOCK 012
Tool no. 1 is indexed into position ready for the next part. The first spindle speed for the job is reset.

BLOCK 013
 Spindle off.

8.7. ARC OFFSETS

N/C lathes supply both linear and circular interpolation. The calculations needed to cut a curve are made by the machine control unit from a single block of information.

There are two restrictions to this capability. No arc can be programmed over more than 90 deg in one information block, and the arc must be entirely in one quadrant. A 120-deg arc could be programmed as two 60-deg arcs if both lie in different quadrants, or two arcs of 90 and 30 deg. For most n/c lathes, the arc radius cannot exceed 9.9999 in.

Circular arcs require the preparatory functions G02 or G03. The convention for clockwise or counterclockwise arcs is the following. The rotational direction of the cutting tool when one is shaping an arc on a horizontal-bed machine is that given by looking upward at the plan view of the work piece (not downward, as is more usual with plan views). For a vertical-bed machine such as the Monarch turN/Center 75, the arc direction is given in Fig. 81.

I and K dimensions were explained in the previous chapter, but are repeated here for convenience. The I offset dimension is the X distance from the starting point of the arc to the center of the arc. See Fig. 43. The minus sign, if any, is ignored. The K offset dimension is the distance in Z from the starting point of the arc to the center of the arc; sign is again disregarded.

Feed rate number for an arc is given by the formula

$$FRN = 10F/R,$$

where R = radius.

8.8. DIRECT FEED RATE

When I and K words as well as X and Z departures are programmed in the information block for G01 straight cuts, the feed-rate number can be entered "direct" as the actual feed rate in ipm, instead of FRN by formula. Consider as an example the 26.5-deg slope cut of Fig. 94. The I word is the sine of the angle and the K word the cosine. Sine 26.5 deg = 0.4462 and cosine = 0.8949. If the feed rate is to be 6 ipm, the information block would read

 X—03 Z06 I04462 K08949 F006

The feed rate is the actual feed rate of 6 ipm. Note also that I and K words are programmed with the decimal point following the first digit.

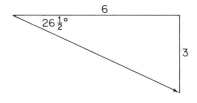

Figure 94

The maximum number that may be coded by this method is 09999. For a cut parallel to the Z axis, I is the sine of 0 deg or zero, and K is the cosine or 1. Since a K of unity cannot be coded, use K09999. Cuts parallel to the X axis are coded I09999 K0, with the feed rate coded direct.

When making a circular cut, the FRN equation must be used. These procedures apply only to straight cuts with function G01. Without I and K words, use the FRN formula.

These procedures are standard for numerically controlled lathes.

8.9. EXAMPLE 2: PROGRAMMING A SEATING PIN

The pin is dimensioned in Fig. 95. It is finish-machined by using only one cutter with a nose radius of 0.031 in. Initial position of the tool is the fully retracted position shown in the figure. A feed of 0.006 in. per revolution is selected. Cutting speed is 500 fpm, giving the following calculated rpm:

For 1-in. diameter: about 2000 rpm. Select S59.
For $4\frac{1}{2}$-in. diameter: about 450 rpm. Select S45.
For 8-in. diameter: about 240 rpm. Select S40.

The feed functions in the program of Fig. 96 are slightly rounded off, since small discrepancies in feed rate have little effect on performance.

The first two columns of the programming sheet should be noted. These keep account of the incremental movements in absolute dimensions. The dimensional accounting in these columns can begin with the actual dimensions of the start position (that is, in the present example, X = 5.7 in.) or with zero. The final total at the bottom of the columns must, of course, give the initial figure; otherwise, there are dimensional errors in the program.

8.10. THREAD CUTTING

Three preparatory functions are used in thread-cutting under numerical control:

G33   thread cutting, constant pitch
G34   thread cutting, decreasing pitch (increasing lead)
G35   thread cutting, increasing pitch (decreasing lead)

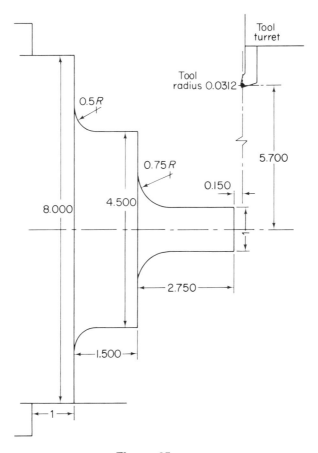

Figure 95

Thermoplastic extrusion machines sometimes require an extruder screw with a variable pitch for processing certain thermoplastic materials. Such a work piece would be produced with a G34 or G35 function. Most n/c lathes, the Monarch turN/Center 75 included, employ only G33.

The "lead" of a thread means "inches per thread," which is the reciprocal of threads per inch. Profile and formulas for the American National Standard thread are given in Fig. 97.

To program a constant-pitch thread, the spindle must first be brought to threading speed in the information block preceding the threading operation, to allow time for acceleration or deceleration to exact threading speed.

MONARCH PATHFINDER LATHE NUMERICAL CONTROL PROGRAM SHEET

OPERATION: _____ DATE: _____ SE.NO.: _____ HOME POSITION: Long. _____ Trans. _____
PLANNER: _____ PART NO.: _____ MATERIAL: _____ STOCK SIZE: _____
CUSTOMER: _____ PART NAME: SEATING PIN HARDNESS: _____ MACHINE: MONARCH 75
TAPE NO.: 0002

ABSOLUTE DIMENSION FROM START		SEQ NO.	PREP. FUNC.	INCREMENTAL DISTANCE		DISTANCE TO ARC CENTER		FEED FUNCTION	SPIN SPEED	TOOL NO.	MISC. FUNC.	OPERATION, TOOL HOLDER, AND TOOL HOLDER LOCATION	SFPM	RPM	DIA.	FEED		DEPTH OF CUT
x axis	z axis			Transverse x±	Longitudinal z±	Parallel to X j	Parallel to Z k	f	s	t	m					IPR	IPM	
		001	04	02					59	11	03	START	500	2000	1	006	12	
0	0	002									08	SHIFT TOOL FOR FACING						
-0.119		003	01		-00119			5				ADVANCE FOR FACING						
-5.05		004		-0505				4				FACE 1" DIA.						
-5.70	-0.019	005		-0065				1846				RETRACT IN Z						
		006			+.001			5				POSITION FOR 1" DIA.						
-5.169		007		+00531				5				TURN 1" DIA.						
	-2.150	008			-02131			057				TURN 3/4 FILLET						
-4.450	-2.869	009	02	+00719	-00719		00119	036	45			FACE 4½ DIA.	450	4½				
-3.419		010	01	+01031	-01031			027				TURN 4½ DIA.						
	-3.900	011						027				TURN ½" FILLET						
-2.950	-4.369	012	02	+00469	-00469		00469	030	40			FACE 8" DIA.	240	8	006	1.44		
-1.669		013	01	+01281				012				TURN 8" DIA.						
	-5.369	014			-.01			0144				RETRACT IN X						
0		015		+01669				5			09	RETRACT IN Z						
	0	016			+05369			37			05							
		017									02							

Fig. 96. Program for seating pin of Figure 95.

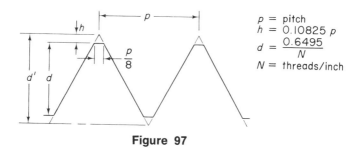

Figure 97

The length of the shaft to be threaded is programmed as Z. If the thread is a tapered thread, then the X dimension must also be programmed. The K word is the thread lead, with decimal point after the first digit as for direct feed-rate programming:

8 threads per inch   K0125
4 threads per inch   K025
6 threads per inch   K01667

The limitation of four decimal places results in a rounding-off error for lead with 6 tpi. For each revolution of thread, the lead is slightly in error, but the error would be significant only if the thread were several feet long. A lathe coded only to three decimal places (thousandths), like most mills and drills, would not be suitable for threading operations.

The diameter of the work piece must be recessed at the end of the threaded portion so that the tool can get into the thread depth or out again. This follows standard practice for engine lathe work.

Usually a thread is formed by using two or more overlapping plunge cuts to reach the final depth of thread. Each plunge cut is offset in Z from the previous cut, as shown in Fig. 98. The Z axis correction is given by

$$Z = X \tan \theta/2$$

Usually θ is 60 deg, so that $Z = 0.5774X$.

8.11. A THREAD-CUTTING EXAMPLE

Figure 99 shows the lathe setup for cutting a 16-pitch thread on a 2-in. shaft. The program is given in Fig. 100.

The depth of cut is 0.0406 in. This depth is programmed in two plunge cuts, the first 0.0206 in., and the second the additional 0.0200 in. To produce the second cut, a Z offset of 0.0111 in. is used.

160 Chap. 8 / *The Numerically Controlled Lathe*

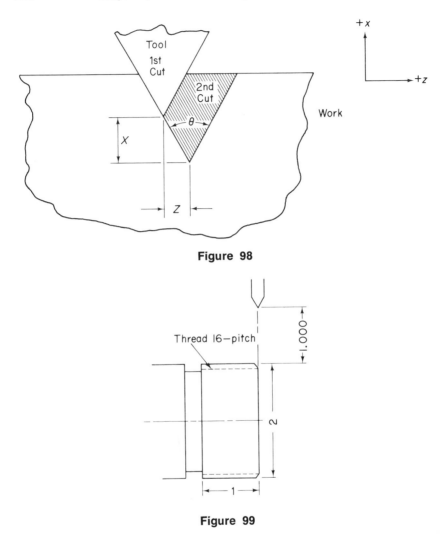

Figure 98

Figure 99

8.12. PRACTICE EXAMPLES FOR N/C LATHE PROGRAMMING

Figure 101 shows a pin to be turned, threaded 10-pitch, and cut off, Fig. 102 an adapter nose, and Fig. 103 a heat sink base. These examples are offered for n/c programming practice.

Manual programming of lathes is not easy, and requires considerable time. Beginners probably should not program the whole series of operations on the part, but single cuts only or all the cuts for one tool only. Program first the coordinates of the tool path. Then complete the pro-

MONARCH PATHFINDER LATHE NUMERICAL CONTROL PROGRAM SHEET

OPERATION: _____ DATE: _____ SE NO.: _____ HOME POSITION: Long._____ Trans._____
PLANNER: _____ PART NO.: _____ MATERIAL: _____ STOCK SIZE: _____
CUSTOMER: _____ PART NAME: 16-PITCH THREAD 2" DIA HARDNESS: _____ MACHINE: _____
TAPE NO.: _____

ABSOLUTE DIMENSION FROM START		SEQ NO.	PREP FUNC.	INCREMENTAL DISTANCE		DISTANCE TO ARC CENTER		FEED FUNCTION	SPIN SPEED	TOOL NO.	MISC FUNC	OPERATION, TOOL HOLDER, AND TOOL HOLDER LOCATION	SFPM	RPM	DIA	FEED		DEPTH OF CUT
x axis	z axis			Transverse	Longitudinal	Parallel to X	Parallel to Z									IPR	IPM	
		n	g	x±	z±	i	k	f	s	t	m							
0		001	04	02					08	11	03	ADVANCE TOOL						
-1.0206		002	01	-010206	-.011	09999	00625	2			08	THREAD FIRST PASS						
-0.7206		003	33			09999	09999	2				RELIEVE TOOL						
		004	01	003	+010589	09999		2				RETURN FOR 2nd PASS						
-1.0406		005			-010889	09999	00625	2				TO THREAD DEPTH						
		006	33									2nd PASS						
0		007				09999	09999	2				RETRACT IN X						
		008	01	+010406	+.011			2			09	RETURN IN Z						
		009	04	02						00	05	CANCEL TOOL OFFSETS						
		010									02							
		011																

Fig. 100. Program for 16-pitch thread.

162 *Chap. 8 / The Numerically Controlled Lathe*

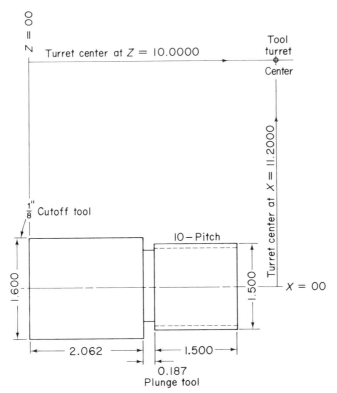

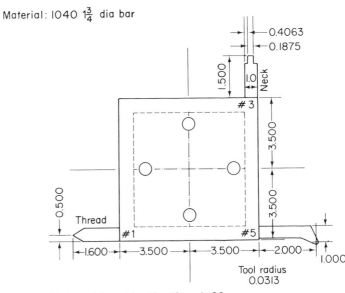

Tool turret layout for Figs. 101 and 102

Figure 101

gramming on manuscript sheets. Keep control of the programming by balancing out the absolute dimensions as recorded in the first two columns of the manuscript sheet. Solutions are offered at the end of the book.

Program the speeds and feed-rate numbers in order to become acquainted with these procedures, but do not check them against those given in the solutions at the back of the book. No one can realistically program speeds and feeds without an acquaintance with the cutters and the lathe to be used. Therefore, any speed and feed is acceptable which is not ridiculously slow or which does not burn up the tool point.

The solution for Fig. 101 produces the 10-pitch thread in two cuts of 0.030 and 0.035 in.

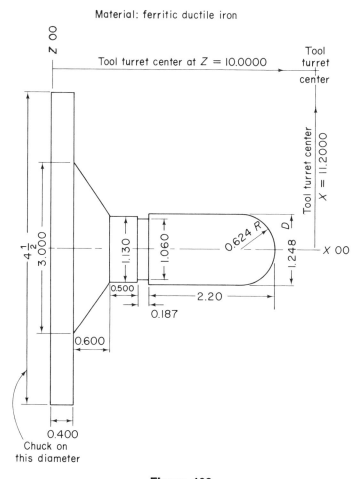

Figure 102

164 *Chap. 8 / The Numerically Controlled Lathe*

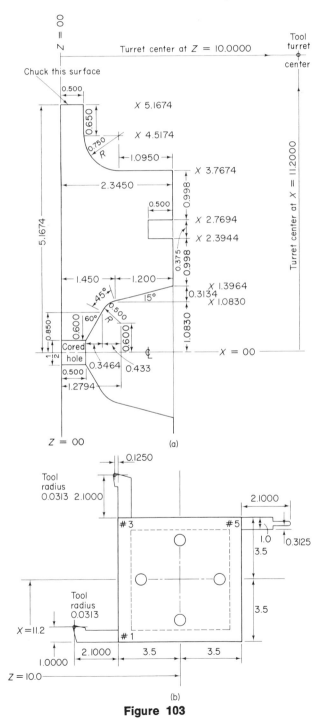

Figure 103

9

THE HARDWARE OF NUMERICAL CONTROL

This chapter describes some of the components of numerically controlled machines and some applications of numerical control. A discussion of the electronic logic elements of the machine control unit cannot be included, however. This is a very extensive subject indeed, and, though it is useful to know, ignorance of it is not a handicap to competence and creativity in the use of numerical control.

9.1. READING THE TAPE

After the work piece, cutting tool, and machine are set up and the tape is located in the tape reader, the CYCLE START or TAPE START button is pressed to begin the numerical control operation. The tape is driven through the tape reader by a low-voltage stepping motor, which receives electrical pulses from a suitable power supply. The tape drive must also be equipped with a brake for fast stopping.

As the tape passes through the tape reader, the holes are sensed. Three methods are in use for hole-sensing:

1. Fluidic (pneumatic)
2. Mechanical
3. Photoelectrical

At least three families of n/c machines, one being the Moog of Chapter 5, read the tape with air in a fluidic method. This method uses the fixed block tape format because each word of information is read at a fixed location in the tape reader. Each block of information, therefore, must have a fixed length so that each word can be positioned under the fluidic tubes that read it. The reading tube terminates at the tape. If there is a hole in the tape, air escapes from the tube and pressure falls. This pressure drop can operate a diaphragm or other air device.

Mechanical tape readers are most common. These are well suited to positioning control. Maximum reading speed is about 60 characters per second, which may be insufficient for contouring operations. Some tape readers use the principle of having pins enter the tape holes to close small electrical switches. A more common kind of reader is that of Fig. 104.

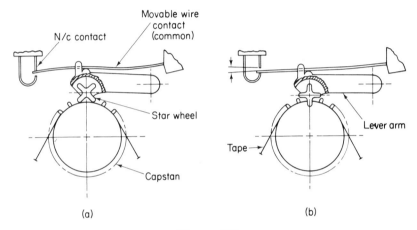

Figure 104

Eight free-rotating star wheels or sprockets ride in parallel on the top of the tape. When one of the points of the wheel drops into a tape hole, the contact attached to the sprocket mechanism opens or closes an electrical circuit. Unlike the fluidic reader, the mechanical tape reader reads rows of characters in sequence.

Photoelectric tape readers are more expensive but have shorter reading times. Their reading speed is usually 300 to 500 characters per second, making them suitable for contouring operations. A light source is positioned on one side of the tape and eight photocells on the other, or nine if the sprocket holes in the tape are to be sensed for tape driving. The presence of a hole gives sufficient light transmission to induce an electric pulse from the photocell. Standard unperforated white paper tape transmits 40 percent of the light from the light source, whereas the hole transmits 100 percent. This difference is sufficient for the electrical characteristics of the photocell.

9.2. PROCESSING THE TAPE INFORMATION

The tape information as read by the reading head is stored in the memory of the machine control unit, each word going to its proper memory position. In a tab sequential system, the first word in the tape goes into the first memory position, the second word into the second, and so on. In a word address system, the word is stored in the memory position called for by the letter address.

When the EOB signal is read, the electronic control logic causes the positioning motors to drive the work table or machine head to the dimensional position called up in the tape. If new positions in both X and Y axes are called for, both motors position simultaneously and usually at the same speed. The resulting motion is, therefore, at 45 deg to both axes until the traverse in one axis is completed, then the motion is parallel to the other axis to the designated point (Fig. 105).

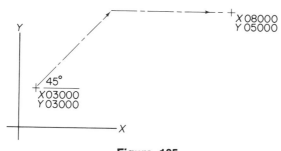

Figure 105

When the required position is reached, the machine control unit acts upon any miscellaneous or preparatory function commands such as tool changes. When the machine has complied with all commands within the block of information, a signal is sent to the tape drive to release the brake, operate the tape drive motor, and to read the next block of information.

This is the sequence of tape reading events for positioning control. It is usually too slow for contouring operations, where tape reading may have to be almost continuous. For contouring, active and buffer storage may be necessary.

Consider the cutting of an arc by linear interpolation with a feed rate of 10 ipm, using a tape reader operating in the method just discussed. The block of information for one cut is read, then executed, and then the next block for the next tangent or secant is read. The feed rate of 10 ipm cannot be maintained over the full length of cut. The machine slides must stop while waiting for the next information block. Therefore, with each block the slides must accelerate up to 10 ipm and at the end of the block decelerate to zero feed. A minor disadvantage is the longer time required to contour. A more serious disadvantage is the poor surface finish. When the slides decelerate, there is a relief of cutter pressure, so that the tool undercuts or gouges at the end of each tangent or secant. If the material of the work piece is an austenitic stainless steel or an exotic aerospace material, there will also be a work-hardening action due to this end-of-block delay, thus reducing tool life.

To improve cutting conditions and speed, the information block may be read into buffer storage at the same time that the machine is operating on the previous block in active storage. At the completion of the machining step, the information in buffer storage is instantaneously transferred to active storage so that the tool can continue cutting without waiting for more information from the tape.

9.3. LINEAR AND CIRCULAR INTERPOLATION

In programming linear interpolation, the beginning of a straight line is programmed in one block of information and the end of the line in the next block. The term "interpolation" means that the machine control unit can obtain intermediate coordinate values lying between the start and the end points of the line.

Linear interpolation is frequently obtained by metering out pulses to the axis drive motors. Suppose that a line must be milled at 30 deg to the X axis. Then in the time interval during which 866 pulses are metered to the X axis drive, 500 pulses must be metered to the Y axis drive. But the feed rate must also be controlled when one is cutting this line. Feed rate is controlled by controlling the rate at which pulses are fed to the two motors.

Circular interpolation capability in the electronics of the machine control unit offers obvious advantages. Since a full 90-deg circular quadrant can be programmed in a single block of information, the operation cannot be delayed by the tape reading operation. There is less programming effort and a better surface finish. The amount of tape information is not increased by a closer tolerance.

Parabolic interpolation is required for certain operations, such as the flame-cutting of ship's plates. The programming and the calculations by the machine control unit are similar to those used in circular interpolation, except that the focus of the parabola is programmed instead of the arc center of a circle.

9.4. OPEN- AND CLOSED-LOOP CONTROL

The axis drive motors used in numerically controlled operations are electric, hydraulic, or electrohydraulic servomotors. A servomechanism is any mechanism used to control displacement, velocity, acceleration, or all of these movements. The design of servomotors is a particularly interesting subject, but it is as complex as that of the solid-state electronics in the machine control unit.

Since the slide mechanism, including the servomotor, must respond rapidly to positional command signals, many mechanical characteristics of the machine tool must be especially designed for numerical control operation. Excessive backlash, for example, would make the positioning devices no more accurate than the backlash would permit. The frictional slideways of conventional machine tools are also undesirable because of their frictional characteristics. Conventional slides have a high coefficient of friction when motion is initiated, this coefficient then decreasing with speed of the slide. Given such a friction performance, motion is usually jerky at low velocities. To obtain a better response, numerical control slide mechanisms use rolling friction from balls and rollers.

In addition to the uniform friction resistance, these improved slide mechanisms require less power. With a smaller power requirement, smaller motors and mechanisms are possible, thus reducing inertia effects. Inertia, however, cannot wholly be controlled by the designer, since it is also at the mercy of the machine user. He may first machine a 1500-lb casting and next a small 2-lb aluminum work piece, thus greatly altering the inertia effects.

In numerically controlled operations, positioning control is obtained by means of either a closed-loop or an open-loop system (Fig. 106). In a closed-loop system, the machine acts on a position command telling the machine where it is to go, while a position sensor reads the position where the machine actually is. A signal from this sensor is fed back to a comparator. The comparator compares the command signal and the position signal and puts out a voltage proportional to the difference between input signals. When the position signal and the command signal coincide, the comparator output is zero and the slide comes to rest. Contouring controls are almost always closed-loop systems.

Open-loop control does not employ feedback. Stepping motors, such as are used in Slo-Syn or Icon numerical control systems, are employed as open-loop units. A stepping motor runs in exact synchronism with its input pulse rate, and is not sensitive to load variations within its capabilities.

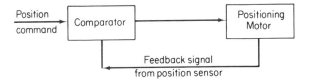

Closed loop system

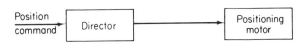

Open loop system

Figure 106

The Slo-Syn and some Fujitsu stepping motors (Fig. 107) rotate 1.8 deg per input pulse. Thus, 200 pulses produce one revolution of the shaft. If the stepping motor drives a lead screw of five threads per inch, then a resolution of 0.001 in. is obtained. Ten threads per inch give a resolution of 0.0005 in. The Slo-Syn motor has a permanent magnet rotor with 50 teeth cut on its periphery. Each pulse produces an angular change

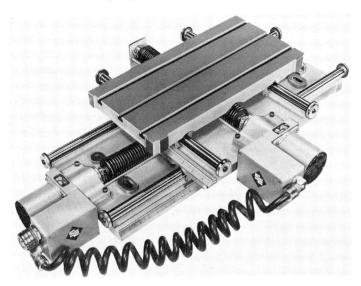

Fig. 107. Servomotor numerically controlled X — Y positioning table "SLO-SYN," both linear motions are controlled by a stepping motor. (Courtesy of Superior Electric Company, Bristol, Conn.)

of one-quarter of a pole, thus giving 200 positions per revolution. Rapid stepping speeds are possible, up to 1000 pulses per second. The stepping motor may be considered as an electrical ratchet.

In addition to driving machine slides, stepping motors are used in the drives for punched tape.

9.5. PROGRAMMING THE SLO-SYN INSTALLATION

In Fig. 108 a Bridgeport mill is fitted with several Slo-Syn motors. X and

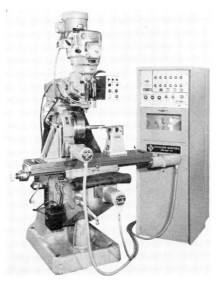

Fig. 108. SLO-SYN stepping motors installed on a Bridgeport mill.

Y motions are controlled, the third axis being a rotating A axis. A helix is being machined on the bar held in the rotating chuck.

The Slo-Syn systems are easy to program. Tape format is tab sequential, with incremental dimensioning in four digits with three decimal places. Each program begins with an EOB and a Rewind Stop character (this is usually the STOP key on a Flexowriter). Both manual and tape-controlled operations are possible. Positioning commands may be inserted manually through digital controls on the control cabinet.

The information block for positioning is the following, where T signifies the TAB key:

Seq. No., T, X, T, Y, T, M, EOB

thus,

3 T 1000 T 3000 T 54 EOB

There are no G functions. The following M functions are employed in positioning control:

- 00 program stop
- 02 end of program and tape rewind
- 06 stop for manual tool change

These are standard M functions. The following, however, are peculiar to the Slo-Syn systems:

- 55 rapid traverse. The system returns in the following block to the machining feed rate set up in the program.
- 56 tool inhibit. Corresponds to G80 of Chapter 3.
- 52 tool advance
- 53 tool retract
- 54 (three-axis machines only) The increment following the second TAB in the block is not a Y dimension, but a command for the third axis, such as the A axis of Fig. 108.

The functions 52 and 53 operate before moving to the position commanded in the same block, not after positioning. These functions also must be coded with positioning commands.

More than one miscellaneous function may be coded in the same block, but 06 and 02 must precede any 50 function. To code two 50's, only the first 5 is needed; thus, 54 with 56 may be punched 546. The combination 02545 means 02, 54, 55.

9.6. POSITIONING PROGRAMS FOR THE SLO-SYN

Three examples of positioning programs are offered in Figs. 109, 110, and 111. The programs begin with an EOB and a Rewind Stop.

FIGURE 109. The tool switch is set to AUTO so that drilling and tool retraction are automatically performed at each XY location on the tape. This tool action is inhibited in the last block with the M function 02. Feed rate is set at HI for rapid traverse. Note the usual sign convention for increments: $+X$ for tool moving to the right, $+Y$ for tool moving away from the operator (up the drawing page).

FIGURE 110. Here a rotary table is used. Function 54 is an instruction to the control to use the second dimension in the block for the third axis, a rotary axis in this case. Each motor step moves the rotary table 0.01 deg, with a maximum rotation in four digits of 99.99 deg. Therefore, in

Positioning Programs for the Slo-Syn 173

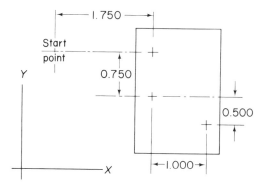

PROGRAM SHEET

PART NO. AND NAME FIG 109						SLO—SYN _____DATE___
SEQ NO.	TAB	X INCR	TAB	Y INCR	TAB	M
Set tool switch to auto for drilling each position						
High	feed rate set					
0						EOB
0	RWS					EOB	Rewind stop
1	T	1750					Drill hole 1
2	T		T	−0750			2
3	T	1000	T	−0500			3
4	T	−2750	T	1250	02	Return to start point

Figure 109

sequence 2 the second dimension is 3000 for a 30-deg clockwise rotation. A negative sign reverses the rotation.

The second hole requires 150 deg of rotation clockwise. This dimension, 15000, is too large to program. It is programmed in sequence 3 as 90 deg and sequence 4 as 60 deg, total 150 deg. However, in sequence 3 a drilled hole is not wanted, so the 56 tool inhibit code must be programmed with 54, as 546.

FIGURE 111. In the two programs above, the tool automatically fed (drilled) and retracted at each XY position when the tool switch was at AUTO. In this program the tool switch is at OFF, and all tool movements will be commanded from the tape.

Sequence 1 positions the tool at the start of the slot, using the 55 hi-feed or rapid traverse code. Sequence 2 mills the slot, using a 52 tool advance. This 52 code goes into effect before positioning to the destination given in the same block.

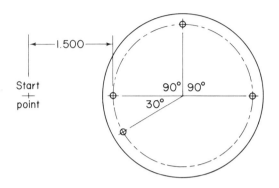

PROGRAM SHEET

PART NO. AND NAME FIG. 110							SLO–SYN _____ DATE _____
SEQ NO.	TAB	X INCR	TAB	Y INCR	TAB	M	
0						EOB	Tool switch at auto
0	RWS					EOB	
1	T	1500					Hole 1
2	T		T	3000	54		2
3	T		T	9000	546		No hole
4	T		T	6000			3
5	T		T	9000			4
6	T	-1500		9000	0254		

Figure 110

Sequence 3 returns the tool to the start point, with three M functions: 02, 53 (tool retract), and 55 (hi-feed). The 53 code is in effect before tool positioning commences.

9.7. CONTOUR MILLING WITH THE SLO-SYN

Contour milling adds little to the complexity of programming a Slo-Syn installation. Both linear and circular interpolation are available in Slo-Syn controls, and the latter, of course, requires the use of I and J words.

A sample program is given in Fig. 112. The M functions are the usual Slo-Syn functions:

55 hi-feed

Contour Milling With the Slo-Syn   175

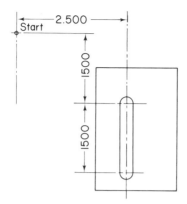

PROGRAM SHEET

PART NO. AND NAME FIG III						SLO—SYN ____DATE____
SEQ NO.	TAB	X INCR	TAB	Y INCR	TAB	M
						EOB
0	RWS					EOB
1	T	2500	T	−1500	55	
2	T		T	−1500	52	
3	T	2500	T	3000	02535	

Figure 111

52 tool advance before positioning
53 tool retract before positioning

Questions

1. Program the O-ring groove of part print no. 1 with Slo-Syn controls. Use a start point 4 in. to the left and 8 in. above the lower edge of the drawing.
2. Program the machining for part print no. 6, calibration block, using Slo-Syn controls. Use a start point 20 mm to the left and 120 mm above the bottom surface of the block. Assume that the stepping motor provides increments of 0.01 mm, with a maximum positioning increment of 99.99 mm. Use a 400-mm end mill and make the convenient assumption that the full thickness of the block can be machined in one cut.

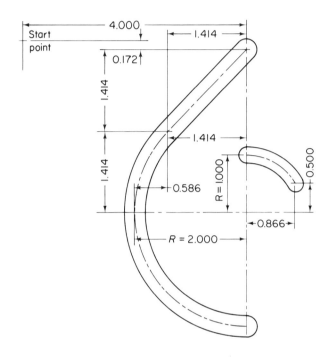

PROGRAM SHEET

PART NO. AND NAME										SLO-SYN DATE
SEQ NO.	TAB	X INCR	TAB	Y INCR	TAB	I	TAB	J	TAB	M
Tool switch at off										
	EOB									
0	RWS	EOB								
1	T	4000	T	0172	T		T		T	55
2	T	−1414	T	−1414	T		T		T	52
3	T	−0586	T	−1414	T	1414	T	1414	T	
4	T	2000	T	2000	T	2000	T		T	
5	T	−4000	T	5000	T		T		T	06535
6	T	4866	T	−2500	T		T		T	55
7	T	−866	T	0500	T	0866	T	0500	T	52
8	T	4000	T	2000	T		T		T	02535

Figure 112

3. Set up a Slo-Syn program to drill all the holes in the base pad, part print no. 2.

9.8. CONTAMINATION OF HYDRAULIC OIL

Most numerically controlled machine tools are operated by hydraulic systems. These systems include pumps, usually piston pumps, spool valves, and servovalves. The mating parts of these hydraulic components are manufactured to accuracies of a few "tenths." A ten-thousandth of an inch is 100 millionths of an inch, which is a smaller dimension than the size of many contaminating particles that enter hydraulic oils.

Hydraulic systems are classified according to required contamination standards as the following:

Class 1—ultraclean system
Class 2—good aerospace missile system
Class 3, 4—critical system
Class 5—poor missile system
Class 6—industrial service

Size and quantity of contaminants allowed in these systems is given in the table below as number of allowable particles per 100 milliliters. The particle size is given in microns, where 1 micron = 0.001 millimeter or $\frac{1}{25000}$ in.

Size Range	Class 1	2	3	4	5	6
5–10μ	4600	9700	24000	32000	87000	128000
10–25μ	1340	2680	5360	10700	21400	42000
25–50μ	210	380	780	1510	3130	6500
50–100μ	28	56	110	225	430	1000
>100μ	3	5	11	21	41	92

A numerically controlled machine should preferably meet the standards of Classes 2 to 4.

Contaminating particles will silt, plug, and cause interference with flow. Probably their worst effect is erosion and wear (the eroded material is a cause of increasing erosion). A single particle is circulated and recirculated at speeds of 20 feet per second and higher, and each time may erode the mating surfaces of critical hydraulic components. The fascinating aspect of wear is the concept that the loss of far less than an ounce of material by wear may reduce a machine to scrap. A 4000-lb automobile costing $4000 may be worth less than $1000 if the engine loses compression because of wear. The loss of weight by wear may be only a half-ounce of material. If the automobile's value fell to $1000 entirely due to wear, then that half-ounce of critical metal lost is worth almost $100,000 a pound. The same considerations, of course, apply to hydraulic systems.

The condition of the hydraulic oil should be carefully guarded by the use of the best maintenance procedures. Oil sampling kits may be used to sample the oil for contamination analysis. The best analyses include a particle count.

The condition of the oil deteriorates with use. New oil added to a system includes many particles in the range 25 to 100 microns. After use over a period of time, these larger particles become reduced in size from attrition, thus increasing the number of abrasive particles. An increased number of abrading particles removes an increasing amount of material from the steel components of the hydraulic system. The further wear proceeds, the faster it proceeds.

Filters are used to control contamination. These are of three types: screen filters, depth filters, and absolute filters. Screen filters are made of woven screens or perforated metal. Depth filters contain a random orientation of material in a mat, so that contaminants of sufficiently large size can be trapped in the complex flow passages of the mat. The only filters that can remove 100 percent of contaminants above a certain size are absolute filters, such as those produced by Millipore Filter Corporation.

Best results are given by a depth filter upstream of an absolute filter. The depth filter traps the bulk of the particles, leaving the absolute filter to trap those particles that manage to escape the depth filter. This method has been highly satisfactory in n/c hydraulic systems.

9.9. TOOLING

Tooling for numerical control operations is a major subject. The basic principles of tool design hold for the special cases of n/c tooling: design for rigidity rather than for high stress level, methods of part location, security and setup, adequate access of the cutting tool to the part, and so on.

Tooling for vertical spindle positioning mills and drills is not usually of complex design. Clamps or milling vises often suffice. But three-axis horizontal mills that must machine several faces of the work piece in a single setup on an indexing table can offer some challenge to the tool designer. A frequent requirement is the machining of two opposite faces of the work piece. This is best handled by the use of picture-frame tooling. Figure 113 is an example. In the middle of the photograph the picture frame is displayed, with clamping and locating components arranged in front of it and a chuck attachment to the left. This is a type of universal

Fig. 113. Components of a picture-frame fixture for milling two sides of a part in numerically controlled milling on a machine as shown in Figure 74. (Courtesy of Kearney & Trecker Corporation, Milwaukee.)

tooling especially designed for the Milwaukee-Matics. The pallet base, to which the tooling is attached, is to the right of the picture frame. The mounting plate for the chuck could be attached upside down to the picture frame, with loss of locational accuracy. Notice that the picture frame and mounting plate are "foolproofed" against this possibility. The three bottom mounting holes on the picture frame and on the mounting plate are not symmetrically laid out; those on the top are.

Security and deformation are major design considerations for this tooling. Accuracy is lost if the picture frame should flex under tool loads; therefore, it is built of heavy sections. Further, the work piece must be adequately secured against being pushed back into the fixture by tool thrust.

Tooling that allows access to several sides of the work piece is expensive, even when made by the n/c machining center itself. The picture frame fixture of Fig. 113 cost more than $1000. Much multi-axis machining in the aerospace industry requires tooling far more expensive than this.

Specifications for perishable tools for numerical control work can be exacting. It is illogical to use a defective drill and drill chuck arrangement with a total runout of 0.005 in. on a machine that can position to 0.001 in.

Drill chucks are rarely used in n/c machining, being replaced with collets or other special holders. Drills must be machine-ground for accurate lip concentricity. Short drills are used so that the drill is as rigid as possible. Spot drills or center drills for n/c drilling should preferably produce a spot a few thousandths larger than the drill to be used. The larger spotted hole guides the periphery of the drill into accurate location rather than the lips of the drill. Spiral-point drills are self-centering and do not require prior spot drilling. Raised markings rolled into the drill shank are another source of inaccuracy.

Since n/c machines are programmed for cutter offset, it is not convenient to regrind worn end mills. Many end mills are a few thousandths of an inch undersize and thus raise the same problems of tool offset.

The problems of tap breakage in numerically controlled machining are generally those encountered in automatic lathe work. Difficulties lie chiefly in the range of $\frac{3}{16}$ to $\frac{5}{16}$-in. diameters. Tap drill sizes to produce 60 percent of full thread are commonly used.

Cutting tools such as drills cannot be casually interchanged between the $50,000 n/c machine and the $200 bench-type drilling machine. The n/c machine must have its own inventory of tools. When these are worn, they can be transferred to cheaper machines doing less demanding work.

9.10. INSPECTION

Though numerically controlled machines can position to 0.001 in. or better, accuracy is dependent, on work piece, tools, tooling, and setup rather, than on control precision. Accuracy may also be influenced by the programming method or even programming errors. The assumption is never made that the tape is correct and therefore so is the part. It is, however, customary in n/c production to inspect only the first and last part of the production run, unless there are special reasons for further inspection.

Suppose a run of 10 pieces is begun, the picture-frame fixture of Fig. 113 being used. To avoid any of the many opportunities for error, it is desirable to finish all 10 pieces without interruption. When the first part comes off the machine, it must be inspected. The machine is set up, ready to produce nine more parts, yet it is risky to make any more pieces until the shop is certain that tape commands are actually producing the part desired, that the fixture is holding the part correctly, and that cutting tools are quite certainly cutting to required depth.

Now inspection requires the use of surface plates, height gages, sine bars, and a multitude of other devices to be manipulated. The inspector must check every hole and dimension and watch for dwell marks and other defects. If the part has contours, these must be examined perhaps on an optical comparator. If traditional methods are used, the part cannot be inspected as fast as the n/c machine can produce it. It may happen that the first part comes off the machine in 12 minutes, but the inspection occupies four hours.

The expensive n/c machine cannot be held in idleness for four hours. If the operator demounts all the fixturing so that he can use the four hours for another production run, then he must set up the fixtures all over again, which means that the second part off must be inspected also.

Fig. 114. Model 300 Sheffield Cordax 3-axis Measuring Machine with computer Printout Accessory System for automatic data handling, n/c tape generation, and mathematical computation.

The realization that inspection was a bottleneck in the numerical-control process came rather slowly. The inspection problem is resolved by the use of numerically controlled inspection machines such as the one illustrated in Fig. 114. The tape used for production may also be used for the inspection. The inspection machine picks off coordinates with a stylus or hole probes and compares actual dimensions with the tape dimension commands, giving a printout of dimension deviations. With computer assistance, such devices can also produce a tape to make a part if given the part to inspect. In the electronics industry, tapes for the machining of printed circuit boards are frequently produced from artwork or drawings, n/c inspection machines being used.

9.11. DRAFTING

Drafting is an obvious application for numerical control. However, n/c drafting has not proved to be economical for routine drafting operations with dimensioned drawings, largely because the accuracy of the n/c drafting machine is not needed. Typical applications lie in large contours, airfoil shapes, automobile bodies, topographical maps, and printed circuit artwork or integrated circuit masks.

The numerically controlled drafting machine is a two-axis machine tool with a large bed and a pen for a cutter. The machine control unit must be equipped for contouring. Very high rates of rapid traverse are employed, up to 600 inches per minute.

Drafting machines are commonly computer-programmed, usually in APT language. Part III of this book devotes limited attention to drafting programs. The APT program used to machine a work piece can be used also to draw the work piece on a numerically controlled drafting board. The same program will also inspect the work piece on an n/c inspection machine.

9.12. FLAME CUTTING AND WELDING

The information processing involved in most welding operations is so limited that there are relatively few numerically controlled welding machines in use. Some electron beam welders, however, have been equipped with numerical control.

Flame cutting of odd or complex shapes is not unusual, especially in such industries as shipbuilding. The operation, therefore, is well suited to numerical control. Computerized programming in a language such as APT may sometimes be used, though often even curved ship's plates are hand-programmed.

Welding and flame cutting require fast tape readers and buffer storage. If the welding gun must wait for the next block of information, the arc will burn a hole in the work piece.

A large European installation for the cutting of ship's plates by numerical control is shown in Fig. 115. This installation has linear, circular, and parabolic interpolation capabilities. Control precision for cutting is 1 millimeter, though heat distortion from the cutting operation has an influence on the accuracy of the work.

9.13. PUNCHING

Punch press operations require relatively expensive tooling. A simple punch to produce a couple of holes in sheet metal may cost a few hundred dollars. If only a relatively few pieces are to be punched, the cost of tooling cannot be justified. Sheet metal panels for automatic control boards, electrical panels, and similar large punched panels are not produced in large runs, and are candidates for numerically controlled production.

Fig. 115. Numerically controlled flame cutting at a shipyard.

The punched steel gasket of part print 8 is an example of a punched part that is more economically produced by numerical control methods if the number required is not large. (Car buffs will find a striking resemblance of this part to the gasket for a former General Motors model.) The part is well suited to an n/c turret punch, which can automatically index the proper punch and die into position for the several sizes and shapes of holes.

Numerically controlled punch presses without tool turret can also produce such parts economically. Quick-change punch and die tooling can be manually changed in about 15 seconds on n/c punch presses.

Positioning control and mechanical tape reading are adequate for n/c punches. The metal sheet is clamped at the back edge and the work table is moved in X and Y axes to the punch position. If a number of small sheets are to be punched, it is more economical to punch all the holes for all the pieces in one large sheet and after punching to cut the sheet into the several small parts. Here again numerical control methods lead to changes for the usual manufacturing sequence.

Contours can be shaped on an n/c punch press by programming a nibbling punch around the contour.

9.14. MACHINE SPECIFICATIONS

The following summary table provides a comparison of the specifications of the numerically controlled machine tools discussed in this book.

9.15. ADAPTIVE CONTROL AND THE INFORMATION SYSTEM

Numerical control is a more systematic method of organizing and using machining or other process information. It has not yet basically changed

	Cintimatic turret drill	Cintimatic vertical mill	Gorton 2-30	MOOG 83-1000MC	K+T Eb	K+T IIIb
Max work piece, lbs			2000		1000	5000
Max X traverse	40 in.	40 in.	30 in.	20 in.	24 in.	38 in.
Max Y traverse	20 in.	20 in.	12 in.	10 in.	16 in.	36 in.
Max Z traverse	25 in.	—	8 in.	5 in.	16 in.	24 in.
Motor hp	5	3	5	$1\frac{1}{2}$	5	10
Machine weight			5050	4500	14,100	
Spindle	vertical	either	vertical	vertical	horizontal	horizontal
Tape format	word add.	word add.	word add.	fixed block nonstandard	word add.	word add.
Positioning	yes	yes	yes	yes	yes	yes
Contouring	optional	optional	yes	no	optional	optional
Tape reading, char/sec	60	60	100	block read	300	300
Dimensioning	absolute	absolute	incremental	absolute	absolute	absolute
Zero shift	yes	yes	yes	±0.020	no	no
Buffer storage	no	no	yes	no	no	no
Axes	3	2	2 or 3	2 + 1	3 or 4	3
MCU	Acramatic	Acramatic	BR3100	MOOG	GE	GE

machining methods, but will do so when adaptive control becomes a more familiar feature of numerical control practice.

A major weakness of numerical control methods at the present time is that these procedures generate none of the data required, but assume that all of the data is available. To obtain his part dimensions, the programmer must read a part print. This is a slow business. We may hope that in the future numerical control, when it becomes a totally effective information system instead of an imperfect conglomerate of components and methods, will be able to supply its own input data. At the other end of the process, the numerically controlled production machine does not check or inspect its own production, even though its inspection "template," the control tape, is available in the tape reader for this purpose.

Some of the required data for the machining process is available only by estimate. We refer here to cutting speeds and feeds. They can be too high or too low. Yet the required data for cutting speeds and feeds lie latent in the work piece material, not in estimating tables. Cutting speed depends, among other factors, on the hardness of the material. This can be found from a hardness test, which is an impression test made on the material by a Brinell, Vickers, or Rockwell indentor. But the cutting tool itself is an indentor. Why should not the cutting tool be supplied with suitable instrumentation continuously to feed back to the control unit information on hardness, so that the speed and feed may be continuously adjusted to material hardness?

This is the principle of adaptive control, which has had limited application to numerical control. Adaptive control picks up data in process from its source, the work piece.

We may expect to see the information system that is numerical control improved by degrees in its scope and effectiveness. The use of a separate machine control unit for each n/c machine is an expense that will in time be superseded by centralized information processing units remote from the machines. Even the punching of tape on electric typewriters must be looked upon as a transitory expedient. The cost and time of computer programming are usually less than those of hand programming. If they are not, then we must conclude either that we have adopted the wrong computers, computer methods, or computer language, or that our information-handling system is at fault.

The remainder of this book is devoted to a description of present techniques in processing numerical control information by computers.

Questions

1. It requires about 7 seconds to change a tool in an n/c vertical spindle mill, whether automatically or manually. Which saves time: repositioning the work table or changing the tool?

2. Using a *positioning* Slo-Syn installation, a milling cut of two straight lines is to be made from 0,0 to 4,4, then to 4,6. Can these two cuts be programmed in a single block of information? Why?
3. Without buffer storage, what tape reading speed would be desirable for a mill contouring in steps of 0.010 in. at 10 ipm? Each block contains a word address for X and Y and five-digit dimensions. Electronic switching times are zero, and the start and braking of the tape reader are assumed to consume zero time.
4. See part print no. 8. Estimate the time to punch all holes on an n/c punch with manual tool change, given the following information.
   The punch must be returned to parking position for a tool change. The following time factors apply:
   a. It requires 1.5 seconds for a move between parking position and sheet metal panel.
   b. Average time to move from one hole to the next with the same punch is 0.25 sec.
   c. Punching time per hole is 0.75 sec.
   d. Tool changing time is 15 sec.
   e. Load-unload time is 15 sec per panel.
5. Consult part print no. 2. Determine the positioning time required for drilling all holes in the base plate (do not include the actual drilling—only positioning). Use a parking position for the tool 6 in. to the left of the bottom of the base pad; the tool must be returned to this parking position for a tool change. Rapid traverse time is 40 ipm. Recall that the machine positions initially at 45 deg to both axes.
6. Estimate the time required to drill the large face of the fluidic manifold, part print no. 3. Rapid traverse is 40 ipm. The parking position of the spindle is 4 in. to the left of the part. Include the following time figures and estimate your own drilling speed and feed:
   a. Setup time, 6 min.
   b. Tape handling, 3 min.
   c. Tool preparation for job, 3 min.
   d. Tear-down time, 3 min.
   e. Tool change, 0.15 min.
   f. Rapid Z time, 0.03 min.

   Feed point is 0.1 in. above the part surface. The spindle must be returned to parking position for a tool change.

APPLICATIONS OF NUMERICAL CONTROL

SCHEDULE OF PART PRINTS FOR PRACTICE IN NUMERICAL CONTROL

These part prints are referred to throughout this book, and may be used for further practice in numerical control programming.

Part Print No. 1. *SMALL GEAR PUMP.* Five drawings.
2. *BASE PAD.* One drawing.
3. *FLUIDIC SCHMITT TRIGGER MANIFOLD.* One drawing.
4. *OFFSET LINK.* Two drawings.
5. *SCROLL CASING FOR A CENTRIFUGAL FAN.* Two drawings.
6. *CALIBRATION BLOCK FOR ULTRASONIC FLAW DETECTION.* One drawing.
7. *FRONT TUBE SHEET, SCOTCH MARINE BOILER.* One drawing.
8. *MANIFOLD GASKET.* One drawing.

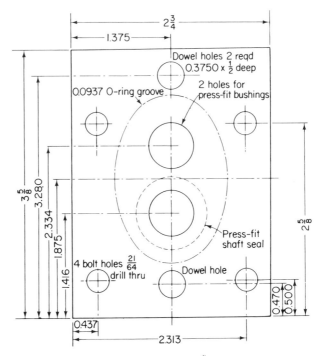

Part Print No. 1. Gear pump drawing no. 1.

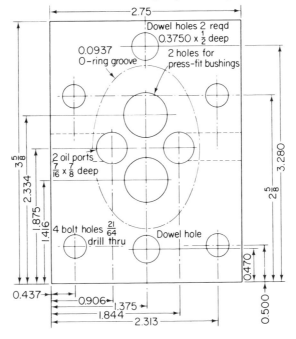

Part Print No. 1. Gear pump drawing no. 2.

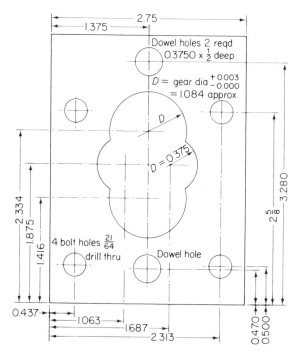

Part Print No. 1. Gear pump drawing no. 3.

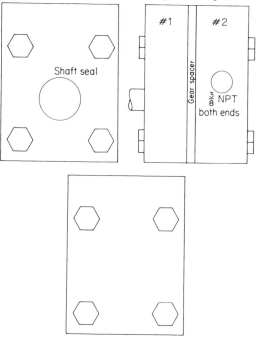

Part Print No. 1. Gear pump drawing no. 4.

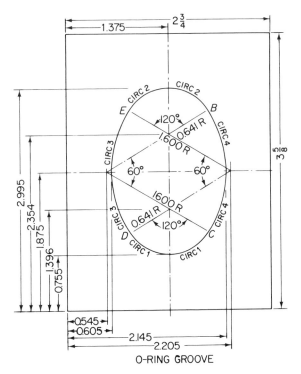

Part Print No. 1. Gear pump drawing no. 5.

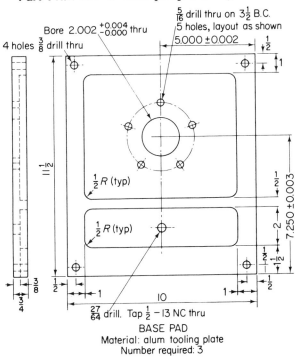

Part Print No. 2

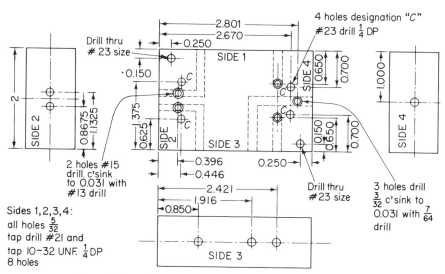

Part Print No. 3. Fluidic Schmitt trigger manifold.

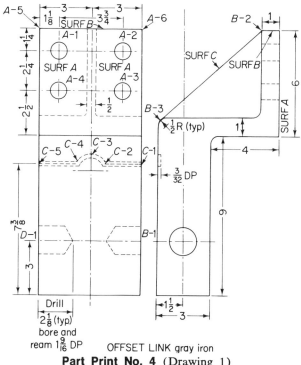

OFFSET LINK gray iron
Part Print No. 4 (Drawing 1)

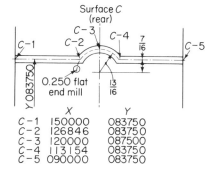

	X	Y
C-1	150000	083750
C-2	126846	083750
C-3	120000	087500
C-4	113154	083750
C-5	090000	083750

Part Print No. 4 (Drawing 2)

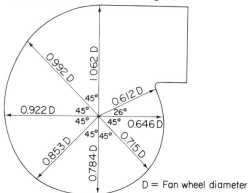

Part Print No. 5. A formula for the scroll casing of a centrifugal ventilating fan. (Drawing no. 1.)

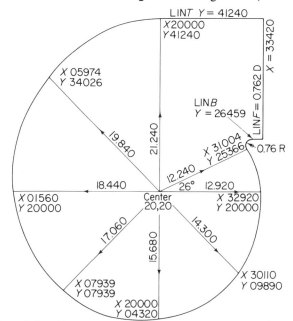

Part Print No. 5. Scroll for 20-inch wheel. (Drawing no. 2.)

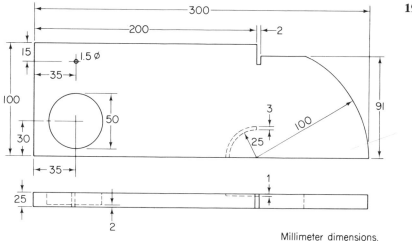

Part Print No. 6. International Institute of Welding calibration block for ultrasonic flaw de-detection.

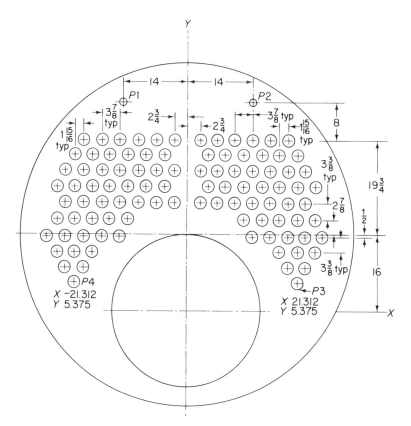

Part Print No. 7. Front tube sheet for a Scotch marine boiler, 72 inches in diameter.

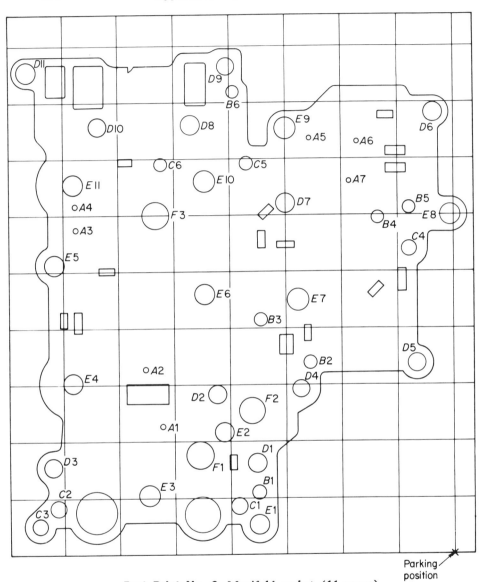

Part Print No. 8. Manifold gasket (11-gauge).

PART 3

COMPUTER PROGRAMMING AND CREATIVITY IN NUMERICAL CONTROL

10

APT: THE BLUE-COLLAR COMPUTER LANGUAGE

10.1. APT AND NUMERICAL CONTROL

Numerical control is a complete technology of information processing. As such, it requires complete facilities for the whole range of numerical-control information processing, including the process machines, the electronic controllers that direct the machines, and, of course, computers and software for processing the information from the raw data to the finished control tape.

The electronic data processing of information by a computer requires the use of a standard computer language for putting the data into the computer, such as Fortran. Fortran has been the popular computer language for technical and scientific work. More recently it has been recognized as an unnecessarily difficult computer language to learn and apply, though it is still useful for some kinds of n/c programming. The computations for the O-ring groove, discussed in Chapter 4, can be handled conveniently with a Fortran program.

Nevertheless, any person learning the practice of numerical control need not learn Fortran. This language requires too much time to learn. Fortunately, there is no lack of computer programmers who are happy to work out a Fortran program for the n/c programmer who needs such help. In any case, Fortran has limited utility for numerical control, and does not serve most of the problems that arise in this type of programming.

There are special computer languages for virtually every technology that can use computers, and numerical control is no exception. There are so many computer languages for numerical control that no purpose is served even in listing the commoner ones, and of course new languages are and will be invented for this purpose. One language, however, is predominant. It has five-axis capability; it can handle almost all the problems that other languages can, and very many that others can not. That computer language is APT.

APT is an acronym for Automatically Programmed Tools. It is a language for the problem of guiding a tool in three-dimensional space. The word "tool" does not restrict APT to cutting tools. The tool may be a drafting pencil, a paint spray nozzle, a cutting torch, a combine harvesting a field of wheat (there is an excellent APT routine for this purpose, but unfortunately no n/c combine), a helicopter making an airborne magnetometer survey, a person looking for a needle in a haystack (a problem in folklore finally solved), or an imaginary tool, or no tool at all if one merely wants the mathematical solution of the program without any ensuing process. But most APT programming is used for production processes; therefore, it has the distinction of being a "blue-collar" computer language.

So that APT might be as comprehensive, as unrestricted, and as convenient as possible, it was developed for large computers. A single statement in APT could exhaust the memory capacity of smaller computers such as the popular IBM 1130. APT programs must be processed on the larger "number crunchers" such as Univac 1108, IBM 7090, or IBM 360/50. These machines rent for $500 to $700 per hour.

These costs would seem to indicate that APT programming is entirely out of reach of the small shop or a training course in n/c programming. But that is too hasty a conclusion. An APT program for the O-ring groove of the gear pump will be discussed presently. If this program is sent to the nearest suitable computer center for processing into a printout of cutter locations, the computing time will be of the order of a few seconds, and the invoice for such service about $10.

So then, APT computer processing is cheap after all. But anyone who has mastered the difficulties of Fortran programming will have another question: how difficult is it to learn APT?

To anyone who has struggled through a course in Fortran, APT programming is surprisingly easy. There are a few qualifications to that statement, however. The APT system has a remarkable range of special techniques; obviously, it requires time to learn all the useful ones, even though each one may be relatively easy to master. There are special techniques or routines for drilling hole patterns, for pocket milling, for putting a curve through a series of empirical points, and so on. Further, it is unusual to get what you want from the computer on the first try, as will presently be shown, and the programmer may have to make minor modifications to his program, perhaps once, perhaps more than once. Finally, of course, programs in five axes are hardly to be described as easy.

APT may be summarized as a most versatile and convenient programming language and system, adaptable to most requirements, not expensive, and not as intellectually demanding as it is sometimes thought to be. The APT programmer does not code the tape, but supplies information to the computer as geometrical definitions. He does not program the computer operations, as he would with Fortran, because the program is built into the APT system. Rather, he supplies enough information for the computer to provide the solution required, and few programmers know how the computer obtains the solution.

10.2. DESCRIPTION OF THE O-RING GROOVE

The O-ring groove, part print no. 1, to be programmed in APT language has been discussed before. It is dimensioned from the lower left-hand corner, this corner being arbitrarily assigned the coordinates X03000 Y10000. The XY plane (Z00000) is taken to be the bottom of the O-ring groove.

The cutter is a ball end mill $\frac{3}{32}$ or 0.09375.

The arcs CIRC1 and CIRC2 have a radius of 0.641 and are both 120-deg arcs. The arcs CIRC3 and CIRC4 have a radius of 1.600 and are 60-deg arcs. The points of tangency of the four arcs are lettered B, C, D, E.

The curve of the quasi-ellipse, of course, represents the center line of the O-ring and its groove. There is a nasty problem in design here. The groove is made quasi-elliptical so that it can enclose the two gears of the small gear pump. Since O-rings are circles and not ellipses, the problem is to make the center line length of the ellipse the same length as the center line of the O-ring selected, which is 2 in. O.D. with all adjacent arcs tangent to each other. The calculations result in the dimensions shown.

10.3. THE APT PROGRAMMING SEQUENCE

The usual sequence of operations for an APT program is this:

1. On the part drawing, lines, planes, circles, and other geometric elements are named. The programmer may select any name of his choice, subject to a very few restrictions. The four CIRC's in the drawing could equally well be named C, CIR, ARK, or ABD. The origin of coordinates is also established for purposes of programming.
2. On programming manuscript paper, the APT program is arranged. The program for the O-ring groove is given in Figs. 116 and 117. The program has three parts.

200 Chap. 10 / Apt: The Blue-collar Computer Language

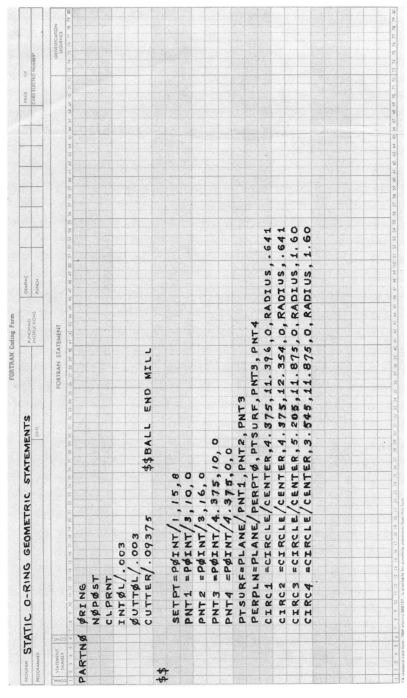

Figure 116

The Apt Programming Sequence

STATIC O-RING TOOL MOVEMENTS

```
FRØM/SETPT
GØTØ/4.437,12.995,8
GØDLTA/0,0,-8              $$RAPID ADVANCE AND SLOW FEED
GØ/ØN,CIRC2,TØ,PTSURF,ØN,PERPLN
GØLFT/CIRC2,TANTØ,CIRC4
GØFWD/CIRC4,TANTØ,CIRC1
GØFWD/CIRC1,TANTØ,CIRC3
GØFWD/CIRC3,TANTØ,CIRC2
GØFWD/CIRC2,ØN,PERPLN
GØDLTA/0,0,8               $$RAPID RETRACT
GØTØ/SETPT

FINI
```

Figure 117

a. Certain general information required for the operation, or to be processed by the postprocessor, such as the name of the part (PARTNO), the diameter of the cutter (CUTTER/ .09375), and others. The cutter diameter is needed so that the computer can calculate the tool offsets from the part shape, though in the case of this groove no tool offset is needed. NOPOST means that no postprocessing is required and only a general solution is asked of the computer. CLPRNT means "cutter location print," an instruction to provide a printout of the many locations of the cutter, as done by hand programming in Chapter 4. Since most n/c machines must approximate circles by linear interpolation, the computer must be told the tolerances to work to. In Chapter 4 both the inside and the outside tolerances were 0.003. This bilateral tolerance is used in this APT program in the statements INTOL and OUTTOL. These first six statements are the general information.

b. A set of geometrical statements that define the geometry of the tool movements. These begin with the statement SETPT = POINT/1, 15, 8. This statement reads, "Setpoint is a point defined by X = 1.0, Y = 15.0, Z = 8.0." The rest of the geometric statements are almost self-explanatory.

c. A set of statements directing the sequence of tool movements. These are given in Fig. 117. Initially the cutter is directed to start from SETPT, from which point it is to go to the point X = 4.375, Y = 12.995, X = 8.000. These statements must be discussed at greater length presently.

3. The program is punched into cards, one statement per card.
4. The program deck of cards is read into the computer, and the required computations are made. The instruction NOPOST means that the programming sequence will end at this point, the computer providing only a cutter location printout.
5. The computed tool positions produced by this first pass through the computer are not suitable for punching into the control tape to operate the n/c machine. What the computer has produced is a general solution to a machining problem. This general solution is deficient in many aspects, most of which cannot be adequately discussed for the moment. But certain things are obvious. If this groove is to be produced on a Cintimatic mill, for example, then the program will require some G78 or G79 preparatory functions. It will require a word address format, and commands to call up the required Z depth. Also, it is common practice to select a convenient origin of coordinates for

the computer program and later to have the computer transform to a different origin of coordinates required by the machine tool, though this is not a requirement in this sample APT program. Therefore, this general solution must be reprocessed through the computer to adapt it to the requirements of the specific machine tool that is to be employed for the work. This machine tool information must be supplied to the computer as a separate item of software called the *postprocessor* program. The postprocessor routine is obtained from the machine tool manufacturer, a computer company, or some other source, and is, of course, applicable to the specific machine only.
6. The final computer output after postprocessing will be given probably on punched cards or magnetic tape, but not punched paper tape, since punching tape consumes too much expensive computer time. The final punched tape will be produced on a magnetic-to-paper tape converter or a card-to-tape converter.
7. The punched tape is inserted into the tape reader of the machine tool. Machining can begin.

10.4. A SURVEY OF THE TOOL MOVEMENTS

As with the geometric statements, it is possible to guess at the meaning of the tool movement commands.

The cutter started at the set point or parking position and went first to 4.375, 12.995, 8. The SETPT must be selected so that the initial tool movement from SETPT will not cause a collision with any holding fixture.

The tool movement commands position the center point of the end mill, that is, the point where the axis of the cutter intersects the bottom plane of the cutter.

The next "go delta" statement is an incremental (delta) movement: the tool moves zero distance in X and Y and -8 in. in Z from where it was in the previous command.

Next a GO ON statement positions the cutter in relation to three planes. To mill this groove, obviously the axis of the cutter must follow the four arcs that form the center line of the groove. The cutter center line must position itself ON CIRC2, that is, without offset, while the end face of the tool is in contact with PTSURF, the plane of the bottom of the groove. To fix the position of the cutter center line uniquely in space, it must also be ON PERPLN, the plane of the vertical axis of the ellipse. By the use of these three planes, the cutter is in position on the ellipse, ready to mill.

The following commands instruct the tool in its path around the ellipse. It must turn left from its previous motion from SETPT to follow CIRC2 until tangent to CIRC4. It follows CIRC4 until it meets the tangent point of CIRC1, then follows CIRC1 to CIRC3, to CIRC2 and back to its starting point, being stopped finally on PERPLN. The tool retracts 8 in. in Z, returns to its SETPT. FINI instructs the computer that this is the end of the program.

10.5. THE PROGRAM PRINTOUT (CLPRNT)

Note the most obvious benefit of this APT program. In APT programming we do not undertake those cumbersome computations for points of tangency of lines with circles or circles with circles or for tool offsets; the APT statements require the computer to do that.

A basic principle of APT programming should be understood at the start. The four CIRC's are *not* arcs. The computer sees them as complete circles, as is apparent from their definitions in Fig. 116, and solves the simultaneous equations required to find the points of tangency.

This statement is not quite true. Actually, the computer conceives of these circles as *cylinders normal to the XY plane*. The APT method is based on surfaces, not lines. Every circle defined as a circle is considered by the computer to be a cylinder; every line is actually a plane normal to the XY plane. In the fourth tool motion statement, the GO/ON statement, the tool is positioned with respect to three surfaces, not two surfaces and a circular line.

This program does not appear to be difficult, either to write or even for the uninitiated to understand. In a sense, we program the tool path we want, but we do not program the computer. This basic characteristic is what makes APT so easy. We virtually take this attitude with the computer: "This is all the information I have (the geometric definitions), and this (the tool movement sequence) is what I want. You are the number cruncher, not I, so you figure it out."

Suppose that we examine the computer printout for the fifth statement of the tool movement part of the program: GØLFT/CIRC2,TANTØ, CIRC4. The statement instructs the computer to program the tool around CIRC2 to the right (i.e., to the right of the part drawing, or to the left as a tool movement) until it is tangent to CIRC4. Here is the CLPRNT for the tool motion command:

X	Y
4.459073997931	12.992493541782
4.621349etc.	12.94896etc.
4.7679—	12.86487—
4.8857—	12.74596—
4.9307—	12.67474—

The last pair of coordinates, 4.9307, 12.6747, is the point of tangency with CIRC4.

The number of points required to program the Cintimatic positioning mill around this arc is 40 to 50, while the above CLPRNT has only five secant points. It is, therefore, suitable only for a contouring mill.

The number of chords required to approximate a circle is approximately equal to

$$\frac{\pi}{2}\sqrt{\frac{D}{E}}$$

where D = diameter of the circle
E = maximum error allowed, which for the O-ring groove was 0.003.

If this formula is applied to CIRC2, the number of chords is 31.

The computer printout gives five points for 60 deg of arc. This corresponds closely with the prediction of the formula.

Can we trust the computer and use these printout values? We can check the results by running the same program through another computer. The set of secant points given was produced by an IBM computer. Here is the printout from a CDC 6600:

X	Y
4.5520189	12.9669505
4.7011336	12.9023438
4.8290886	12.8021602
4.9301804	12.6743993

Compare the two printouts. The final pair of points giving the point of tangency with CIRC4 is 4.930 and 12.674 by both computers. But there is a difference of two-thirds of a thousandth between the two X values for this point. This might be significant if the machining were to "tenths." The discrepancy lies in the programming; it is accurate only to thousandths, and neither computer can upgrade this level of accuracy. If the IBM printout is used, the point of tangency will be punched into the tape as

X04931 Y12675

and the CDC 6600 figures as

X04930 Y12674

after rounding off.

The CDC 6600 printout offers only four points; the IBM printout five. And, of course, the points do not correspond, except the last one. This suggests that one of the computers is in error. Both have given correct figures, however. The IBM program begins with a full secant, and the CDC printout begins with a half-secant. Such minor differences in processing must be expected.

The IBM printout provides 35 points in going all around the O-ring groove; the other printout 28. More points are obtained by tightening the tolerance. Here is the CDC 6600 printout for the same half-arc with a tolerance of ±0.001:

X	Y
4.4782296	12.9856199
4.5699733	12.9635781
4.6574792	12.9282870
4.7388453	12.8805136
4.8123032	12.8212964
4.8762562	12.7519224
4.9301804	12.6743993

There are now seven points instead of four, even though the tolerance was reduced two-thirds.

In order to obtain a printout with values that can be hand-punched into a tape for the Cintimatic mill of Chapter 3, many more points than seven are required for the tape. A tolerance of a few tenths will provide the required number of points for such an XY positioning mill. If we wish the computer to program the whole tape including G and M functions and sequence numbers, we must run this computer general solution through the computer again, using some additional postprocessor commands. One of these postprocessor commands must tell the computer which machine tool-machine control unit is to be used. For the Cintimatic positioning vertical mill, the machine is identified by the statement

MACHIN/CINMAT,1.

The statement NØPØST must be removed from the program if postprocessing is required.

This has been an introductory discussion of APT programming and some of its simpler manipulations. It is an interesting computer routine and generally not difficult. Yet, like all skills, it has its special manipulations, which must be learned by experience. It is possible to write a "perfect" APT program that will not be accepted by the computer. Such cases are instructive, and one such case will be examined in a later chapter.

10.6. PREPARING THE APT MANUSCRIPT

Most APT programming begins with a conventional engineering drawing. Conventional drafting practice is not generally suited to APT programming, and the programmer usually must mark the drawing for ease of programming.

STEP 1. SYMBOLIC NAMES FOR GEOMETRIC ELEMENTS. The programming of the O-ring groove requires reference to four circles. To avoid confusion between circles, the programmer marks each circle with the name he selects for it. This, then, is the first step in programming: Identify and mark the points, lines, circles, planes, and other geometrical elements.

The choice of geometric names is the programmer's choice. On the drawing of the groove, the circles are named CIRC1, CIRC2, and so on. They could be named CIR, CI, C, CIRQ, XYZ, Q1A1, or almost any other choice. The only restrictions on the choice of name are these:

1. The identifying name must not have more than six characters.
2. The name may be mixed letters and numbers, but cannot be wholly numbers. There must be at least one letter.
3. Symbols such as decimal point, comma, equation sign, which are neither numbers nor letters, are not permitted.
4. The selected name must not be a word in the APT computer vocabulary. A complete APT vocabulary is not offered in this book; nevertheless, the beginning programmer will have no difficulty with this limitation. In brief, identifying names such as the following, and others to be found in the programming information which follows, are not allowed:

CIRCLE	PLANE	INTOL
POINT	CUTTER	END
CENTER	PERPTO	ON
RADIUS	TANTO	PARLEL
LINE	FINI	PARTNO

STEP 2. SELECTION OF AN ORIGIN OF COORDINATES. In the part print for the O-ring groove, the origin of coordinates on the drawing is the lower left-hand corner of the part. In the corresponding APT program, this corner is given the coordinates $X = 3$, $Y = 10$. The part was then set up on a Cintimatic no. 3 mill with this corner located at $X = 3$, $Y = 10$.

APT programming can be used to shift or even rotate coordinate axes. The programmer can use any convenient point on the drawing for his origin of coordinates and can set up any convenient axes, without for the moment concerning himself with the location of the coordinate system on the n/c machine. The center of the O-ring groove would be entirely suitable as an origin, for example.

STEP 3. LAYOUT OF THE MANUSCRIPT SHEET. The manuscript paper for an APT program is a standard computer sheet. The punched card has 80 vertical columns; so has the manuscript paper. If the word CIRC1 is printed

>   C in column 23
>   I in column 24
>   R in column 25
>   C in column 26
>   1 in column 27

the keypunch operator will punch these characters into the same columns on the punched card.

No part of the APT program may be written in columns 73 to 80. These columns are reserved for statement numbers, which are also card numbers.

Using the geometric statements of Fig. 116 as an example, we see that PARTNO ORING is the first statement. On this line containing the PARTNO statement, a 1 might be entered in column 80. The next statement is NOPOST, perhaps identified as card (statement) 2 in column 80. The keypunch operator punches each APT statement—each line of the APT program—into a separate card, using the numbers recorded in columns 73 to 80 as card numbers.

However, it is not usual to number the APT statements consecutively. Few APT programs get through the computer without the need for a few modifications, and they may require extra cards. Statement numbers could be 1, 2, 3, 5, 10, 15, or perhaps 10, 20, 30, 40, 50.

The APT words PARTNO and FINI should be placed in columns 1 to 6. Other statements can be arranged on the line as the programmer pleases. He can leave spaces (empty columns) as he sees fit, for neatness, legibility, or any other reason.

In ordinary printing, words are separated by spaces. In APT programming, words are separated by commas as follows:

>   NOY, SPLINE,3,0.825,6.5,0.999,9.4,1.043, etc.

On completion of the manuscript, the statements are punched into cards. The keypunch operator is not usually the programmer, and the punch operator must be able to read the manuscript without any possibility of misunderstanding exactly what each character is. The manuscript, therefore, must be printed with care, in capital letters, one character per column (unless spaces are used). Commas and decimal points especially must be clearly differentiated, and mild disasters result if they are not. Print the letter O as Ø, the number zero as 0. Print the letter I as I (crossed top and bottom), the number 1 as 1 (crossed bottom, inclined top).

If the keypunch operator cannot read your script, do not expect him or her constantly to refer back to you for interpretation. The keypunch job just does not lend itself to this kind of leisurely conference method. If the operator is in doubt, she will take a chance and punch what seems best. Consider what may happen to a slovenly manuscript.

3,0.825,6.5,0.999. These are two pairs of XY coordinates. Suppose that the keypunch operator cannot tell your commas from decimal points. So she punches the information

3.0,825,6.5,0.999.

Finally, it is most convenient to set up the geometric statements (the information input) on one sheet, and the tool motion statements (the desired output) on another. It is then possible to work back and forth between the two parts of the program.

10.7. GENERAL INFORMATION STATEMENTS

The general information statements of Fig. 116 will first be reviewed here.

1. PARTNO. This is the title of the program and the name of the part. There are no limitations on the length of the name or the number of words. The PARTNO is, of course, no part of the APT programming, but is merely a title block that is passed through the computer programming from input to output. NOPOST. This statement is discussed below.
2. CLPRNT. Requests a printout of positions of the axis of the cutter. In the O-ring groove example, this CLPRNT is the actual shape of the groove (within tolerance), since the cutter is not offset.
3. INTOL/0.003. Instructs the computer that the tool may gouge inside the line of the O-ring groove by a maximum of 0.003.
4. OUTTOL/0.003. Instructs the computer that the tool may have a maximum offset from the line of the O-ring groove of 0.003.
5. CUTTER/0.0937. Defines the diameter of the end mill. The computer uses this information to calculate how much the tool must be offset from the work to give the required part shape and dimensions. In the present case there is no offset; therefore, the computer would supply the same set of points in the CLPRNT for CUTTER/0, CUTTER/0.0937, or any other diameter.
    In numerically controlled drafting, the drafting pen, as for any drafting, draws a line without offset from the line. In n/c drafting, then, the statement CUTTER/0 is used.
6. FINI. This is the last card in the APT program, advising the computer that the program is terminated.

A few other general information statements should be mentioned.

7. MACHIN/xxxxxx,n. This statement tells the computer which postprocessor program (which n/c machine) is to be used. After the slash, the name of the machine is inserted, CINMAT for example, followed by a comma, then a number. If the CLPRNT is to be postprocessed into a punched tape, then NOPOST is not included in the program.

8. NOPOST. Tells the computer that no postprocessor or tape will be produced and that only a CLPRNT is desired.

9. TOLER/.nnnn. If no INTOL or OUTTOL statement is included in the program, the computer will use the following tolerances: intol 0.0000, outtol 0.0005. Instead of INTOL and OUTTOL, the statement TOLER/ may be used, such as TOLER/0.00015. The computer is programmed to use this TOLER as an outtol with zero intol. Or if INTOL and OUTTOL are used, and later a different tolerance is required for a part of the program, TOLER/ can be inserted in the proper place in the tool motion statements to override the previous tolerance instructions. The statement TOLER/ will hold until countermanded by another tolerance command.

10. MCHTOL/.nnn. Suppose that the n/c machine must run the cutter around a sharp 90-deg corner such as that shown in Fig. 118. The work table and work piece may have a combined

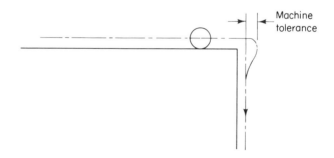

Figure 118

weight of several thousand pounds, and their inertia will cause the cutter to follow the path shown. The MCHTOL/ statement instructs the computer as to how much machine inertia tolerance will be allowed in these circumstances. The computer controls this dynamic tolerance by programming a deceleration in feed rate during the postprocessing.

11. *Other Postprocessor Statements.* To set up an APT program that will produce the final punched tape requires that the programmer call up coolant, spindle rotation, canned cycles such as drilling, and many others, all to be put into the program during the postprocessor stage. These statements may vary with the n/c machine used and the requirements of the postprocessor routine. Typical statements of this kind are

>FEDRAT/20
>SPINDL/900,CLW
>COOLNT/FLOOD

The postprocessor manual and postprocessor tapes are required to complete these instructions.

12. REMARK. Any nonprocessing comments may be inserted into the APT program as statements preceded by the word REMARK. The REMARK may be on a separate card or attached to an APT statement, as the following REMARK is.

>CUTTER/0.09375   REMARK BALL END MILL

Instead of REMARK, the double dollar sign may be used for remarks:

>CUTTER/0.09375 $$BALL END MILL

13. $. The single dollar sign is used at the end of a line when an APT statement must be continued on another line. This symbol thus indicates that the information on the following line (following card) is a continuation of the statement and not another statement.

10.8. THE STRUCTURE OF APT GEOMETRIC STATEMENTS

In Fig. 116, following the few general information statements, the geometric statements are listed. These have the same construction:

>(name of geometric element) = (type of element)/specific details

Thus

>CIRC1 = CIRCLE/CENTER, 4.375,11.396,0,RADIUS,0.641

This is read "CIRC1 is the circle with center at $X = 4.375$, $Y = 11.396$, $Z = 0$, with radius 0.641."

Four circles are defined in the same way, and four points are defined by their X, Y, Z coordinates, always in the order X, Y, Z. PTSURF is a plane defined by the three points it passes through. PERPLN is defined as a plane perpendicular to PTSURF.

Whereas the name of the element is the programmer's choice, the type of element is an APT vocabulary word that must not be misspelled. If PLANE is inadvertently spelled PLAN, or PLANES, the computer will not operate on the statement information. After the slash, the defining information unique to the geometric element is provided in a fixed order that must not be varied. Only certain formats for the defining information are allowed, and the programmer is not at liberty to invent new ones. The following chapter supplies some of these allowed formats.

Note that some of the definitions in the APT program under discussion depend upon prior definitions. PTSURF is defined in terms of three points PNT1, PNT2, PNT3. Therefore, the three PNT's must first be defined before PTSURF is defined; otherwise, the computer is helpless.

A very basic difference between the drawing and the APT program must be noted. Anyone reading the drawing of the O-ring groove sees CIRC1 as a 120-deg arc, not a circle. Since in the APT program CIRC1 is defined as a circle, the computer sees this as a full 360-deg circle. Similarly, the computer sees the two defined planes as extending to infinity in all directions. The programmer must at all times see these geometric elements as he defined them, not as they are drawn; otherwise, he may be unpleasantly surprised. Figure 119 suggests the sort of thing that might

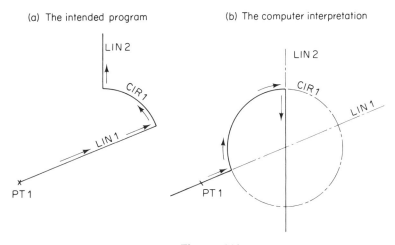

Figure 119

happen. The programmer wishes the tool to start from PT1, follow LIN1 to CIR1 and then follow LIN2. The computer will see an additional intersection of LIN1 with CIR1, and the output of the computer will be the path shown.

The computer sees the geometric statements for the O-ring groove as drawn in Fig. 120. To repeat, all circles are cylinders normal to the

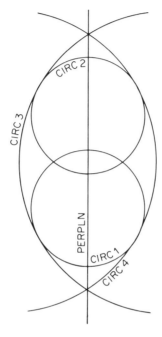

Figure 120

XY plane, and all lines are planes normal to the XY plane.

10.9. THE APT PROCESSOR SYSTEM

Processing of the APT program requires several computer stages.

Section 1, or the first pass of the computer processing, translates part program statements and reduces geometric statements to a certain mathematical definition for each type of geometry known as the canonical form. The canonical form of geometric statements will not be discussed in this book, but no doubt the beginning APT programmer will encounter this term so often that he will be driven by his curiosity to look it up.

Section 2 calculates the coordinates for the consecutive locations of the tool end.

Section 3 prints out the CLPRNT.

Section 4 is the postprocessor section.

Questions

1. Why will the same CLPRNT result from the O-ring groove program if the cutter diameter is changed?
2. Approximately how many chords will approximate a full circle with a diameter of 3.200 in. and a tolerance of
   a. ±0.003?
   b. ±0.001?
   c. ±0.0001?
3. If no INTØL or ØUTTØL statement is included in the APT program, what tolerances are used in the computer routine?
4. Why is a postprocessor necessary for programming an n/c machine by the APT method?
5. Explain MCHTØL.
6. What do the following statements probably mean?
   FEDRAT/6
   SPINDL/600,CLW
7. What Cintimatic G function probably is indicated by the postprocessor statement CYCLE/DRILL?
8. With a suitable program it is possible for a computer to output a punched tape for a specific n/c machine tool-machine control unit without the postprocessing operation. What are the advantages and disadvantages of such a procedure?

11

APT: GEOMETRIC DEFINITIONS

This chapter is a partial dictionary of APT geometric definitions. The beginning programmer should carefully examine the formats of the first few definitions for LINE, CIRCLE, etc. and skim rapidly through the others, which can be examined more carefully as needed.

A word of caution. Do not memorize any APT definitions. Memory is deceitful and can cause needless trouble in programming. You can always consult this or other manuals to check the accuracy of your definitions. Every skilled programmer does—that is why he is skilled.

A Partial List of Useful Definitions

These are explained within this chapter.

PØINT Definitions

1. PT = PØINT/X,Y,Z
2. PT = PØINT/X,Y
3. PT = PØINT/CENTER,(circle)
4. PT = PØINT/INTØF,(line),(line)

LINE Definitions

1. LIN = LINE/X_1,Y_1,X_2,Y_2
2. LIN = LINE/X_1,Y_1,Z_1,X_2,Y_2,Z_2
3. LIN = LINE/(point),(point)

4. LIN = LINE/(point),$\genfrac{}{}{0pt}{}{\text{LEFT}}{\text{RIGHT}}$,TANTØ,(circle)

5. LIN = LINE/$\genfrac{}{}{0pt}{}{\text{LEFT}}{\text{RIGHT}}$,TANTØ,(circle),$\genfrac{}{}{0pt}{}{\text{LEFT}}{\text{RIGHT}}$,TANTØ, (circle)

6. LIN = LINE/(point),ATANGL,degrees

7. LIN = LINE/PARLEL,(line),$\genfrac{}{}{0pt}{}{\genfrac{}{}{0pt}{}{\text{XLARGE}}{\text{XSMALL}}}{\genfrac{}{}{0pt}{}{\text{YLARGE}}{\text{YSMALL}}}$,offset

PLANE Definitions

1. PL = PLANE/(point),(point),(point)
2. PL = PLANE/(point),PARLEL,(plane)
3. PL = PLANE/PARLEL,(plane),$\genfrac{}{}{0pt}{}{\genfrac{}{}{0pt}{}{\text{XLARGE}}{\text{YLARGE}}}{\genfrac{}{}{0pt}{}{\text{ZLARGE}}{\text{XSMALL}}}$,offset
etc.
4. PL = PLANE/PERPTØ,(plane),(point),(point)

CIRCLE Definitions

1. CIR = CIRCLE/X,Y,Z,radius
2. CIR = CIRCLE/X,Y,radius
3. CIR = CIRCLE/CENTER,(point),RADIUS,radius
4. CIR = CIRCLE/CENTER,(point),TANTØ,(line)
5. CIR = CIRCLE/CENTER,(point),(point)
6. CIR = CIRCLE/CENTER,(point),$\genfrac{}{}{0pt}{}{\text{LARGE}}{\text{SMALL}}$,TANTØ,(circle)
7. CIR = CIRCLE/$\genfrac{}{}{0pt}{}{\genfrac{}{}{0pt}{}{\text{XLARGE}}{\text{XSMALL}}}{\genfrac{}{}{0pt}{}{\text{YLARGE}}{\text{YSMALL}}}$,(line),$\genfrac{}{}{0pt}{}{\genfrac{}{}{0pt}{}{\text{XLARGE}}{\text{XSMALL}}}{\genfrac{}{}{0pt}{}{\text{YLARGE}}{\text{YSMALL}}}$,(line),$
RADIUS,radius
8. CIR = CIRCLE/$\genfrac{}{}{0pt}{}{\genfrac{}{}{0pt}{}{\text{XLARGE}}{\text{XSMALL}}}{\genfrac{}{}{0pt}{}{\text{YLARGE}}{\text{YSMALL}}}$,(line),$\genfrac{}{}{0pt}{}{\genfrac{}{}{0pt}{}{\text{XLARGE}}{\text{XSMALL}}}{\genfrac{}{}{0pt}{}{\text{YLARGE}}{\text{YSMALL}}}$,$\genfrac{}{}{0pt}{}{\text{IN}}{\text{OUT}}$,(circle),$
RADIUS,radius
9. CIR = CIRCLE/$\genfrac{}{}{0pt}{}{\genfrac{}{}{0pt}{}{\text{XLARGE}}{\text{XSMALL}}}{\genfrac{}{}{0pt}{}{\text{YLARGE}}{\text{YSMALL}}}$,$\genfrac{}{}{0pt}{}{\text{IN}}{\text{OUT}}$,(circle),$\genfrac{}{}{0pt}{}{\text{IN}}{\text{OUT}}$,(circle),$
RADIUS,radius

CYLNDR Definitions (cylinders)

1. CYL = CYLNDR/(point),a,b,c,radius

TABCYL, GCØNIC and QADRIC

These are especially useful definitions. Their discussion is deferred until later.

11.1. GENERAL CONCEPTS FOR SURFACE DEFINITIONS

1. APT geometric definitions, with the single exception of points, are definitions of surfaces. A line is a plane perpendicular to the XY plane, and a circle is to be considered a cylinder normal to the XY plane.
2. Modifier words LEFT and RIGHT. See Fig. 124. L1 is RIGHT with respect to CIRC1, while L2 is LEFT. The hand is fixed by looking from the point toward the circle.
3. Modifier words XLARGE, etc. See Fig. 128. L1 is YLARGE or XSMALL with respect to L2. L2 is XLARGE or YSMALL with respect to L1, that is, in the direction of larger X values or smaller Y values.

11.2. PØINT DEFINITIONS

Points may be defined by their X, Y, Z or X, Y coordinates. A point may also be defined as the center of a circle (Fig. 121).

$$P1 = PØINT/CENTER,CIRC1$$

or as the intersection of two lines (Fig. 122)

$$P1 = PØINT/INTØF,L1,L2$$

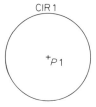

Figure 121

11.3. LINE DEFINITIONS

1. By the coordinates of two points on the line (Fig. 123).

$$L1 = LINE/1,3,5,5$$

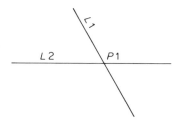

Figure 122

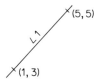

Figure 123

2. Or with Z coordinates

    L2 = LINE/3,3,2,−4,6,1

3. By symbolic names of two points on the line

    L3 = LINE/PT2,PT4

4. Line defined by a point and tangency to a circle. There are two possible tangencies to the circle by lines from the point. These are differentiated by the modifiers RIGHT and LEFT. See Fig. 124.

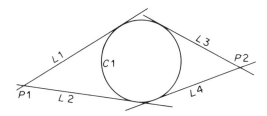

Figure 124

    L1 = LINE/P1,LEFT,TANTØ,C1
    L2 = LINE/P1,RIGHT,TANTØ,C1
    L3 = LINE/P2,RIGHT,TANTØ,C1
    L4 = LINE/P2,LEFT,TANTØ,C1

5. Line tangent to two circles. Here four geometrical possibilities must be differentiated. See Fig. 125.

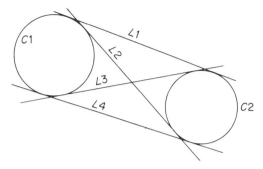

Figure 125

L1 = LINE/LEFT,TANTØ,C1,LEFT,TANTØ,C2
or L1 = LINE/RIGHT,TANTØ,C2,RIGHT,TANTØ,C1

L2 = LINE/LEFT,TANTØ,C1,RIGHT,TANTØ,C2
or L2 = LINE/LEFT,TANTØ,C2,RIGHT,TANTØ,C1

L3 = LINE/RIGHT,TANTØ,C1,LEFT,TANTØ,C2
or L3 = LINE/RIGHT,TANTØ,C2,LEFT,TANTØ,C1

L4 = LINE/RIGHT,TANTØ,C1,RIGHT,TANTØ,C2
or L4 = LINE/LEFT,TANTØ,C2,LEFT,TANTØ,C1

LEFT and RIGHT are as viewed from the first circle specified in the definition to the second circle specified.

6. Line through a point at an angle to the X axis (Fig. 126).

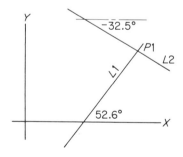

Figure 126

L1 = LINE/P1,ATANGL,52.6
L2 = LINE/P1,ATANGL,−32.5

A line may also be defined by a point and an angle with a specified line (Fig. 127).

L1 = LINE/P1,ATANGL,85.1,L2

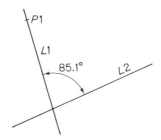

Figure 127

Angles are positive if measured counterclockwise and negative if clockwise. In the first two of the above definitions the angle to the X axis was implied in the definition. Angles must be designated in degrees and decimal degrees.

7. Line parallel to and offset from a given line (Fig. 128).

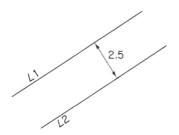

Figure 128

```
    L1 = LINE/PARLEL,L2,YLARGE,2.5
    L2 = LINE/PARLEL,L1,YSMALL,2.5
 or L2 = LINE/PARLEL,L1,XLARGE,2.5
    L1 = LINE/PARLEL,L2,XSMALL,2.5
```

The offset distance is measured normal to the planes. It is not an X or Y component.

A line may also be defined by a point and parallelism (Fig. 129).

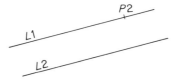

Figure 129

L1 = LINE/P2,PARLEL,L2

11.4. PLANE DEFINITIONS

1. Plane passing through three points not all on the same straight line.

    PL = PLANE/P1,P2,P3

2. Plane through a point parallel to another plane (Fig. 130).

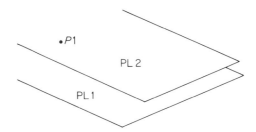

Figure 130

PL2 = PLANE/P1,PARLEL,PL1

3. Plane parallel to another plane and offset from it (Fig. 131).

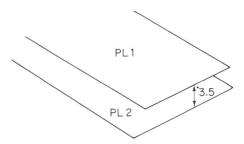

Figure 131

PL1 = PLANE/PARLEL,PL2,ZLARGE,3.5

    The offset distance is measured perpendicular to the planes.
4. Plane perpendicular to another plane and passing through two points, neither point being on the intersection line of the planes.

    PL1 = PLANE/PERPTØ,PL2,P1,P2

11.5. CIRCLE DEFINITIONS

1. Circle defined by the coordinates of the center and the length of the radius.

$$C1 = CIRCLE/2,3,0,4.437$$

2. Or, in two dimensions, X and Y.

$$C1 = CIRCLE/2,3,4.437$$

3. Or by the symbol for the center point.

$$C1 = CIRCLE/CENTER,PT3,RADIUS,4.437$$
$$\text{Also, } C1 = CIRCLE/CENTER,2,3,0,RADIUS,4.437$$

4. By the center point and a line tangent to the circle (Fig. 132).

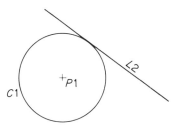

Figure 132

$$C1 = CIRCLE/CENTER,P1,TANTØ,L1$$

5. By a center point and a point on the circumference (Fig. 133).

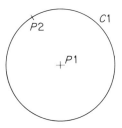

Figure 133

$$C1 = CIRCLE/CENTER,P1,P2$$

6. By the center point and tangency to another circle (Fig. 134).

There are two geometrical possibilities. The modifiers SMALL and LARGE indicate whether the selected circle has the smaller or the larger radius that gives tangency to the other circle.

$$C1 = CIRCLE/CENTER,PT1,SMALL,TANTØ,C3$$
$$C2 = CIRCLE/CENTER,PT1,LARGE,TANTØ,C3$$

7. By a radius and tangency to two intersecting lines (Fig. 135).

Circle Definitions

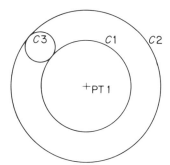

Figure 134

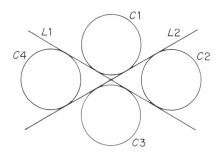

Figure 135

C1 = CIRCLE/YLARGE,L2,YLARGE,L1,RADIUS,.75
C2 = CIRCLE/YSMALL,L2,YLARGE,L1,RADIUS,.75
C3 = CIRCLE/YSMALL,L1,YSMALL,L2,RADIUS,.75
C4 = CIRCLE/YLARGE,L2,YSMALL,L1,RADIUS,.75

Either line may be mentioned first in the definition. The YSMALL and similar modifiers indicate the side of the line that would give the center of the circle the smaller or larger X or Y value of the two possible values. The SMALL and LARGE modifiers often may be defined in either X or Y.

8. By a radius and tangency to a line and a circle (Fig. 136). This case arises when one is fairing a line into a circle by means of a small fillet.

C1 = CIRCLE/YLARGE,L1,XLARGE,OUT,CIR,RADIUS,0.5
C2 = CIRCLE/YSMALL,L1,XLARGE,IN,CIR,RADIUS,.5
C8 = CIRCLE/YSMALL,L1,XSMALL,IN,CIR,RADIUS,.5

Compare the definitions for C2 and C8. Both circles are YSMALL to L1, and both are IN the circle CIR. To differenti-

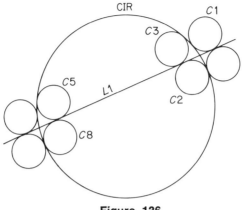

Figure 136

ate between them, C2 is the XLARGE and C8 the XSMALL possibility.

C3 = CIRCLE/YLARGE,L1,XLARGE,IN,CIR,RADIUS,.5
C5 = CIRCLE/YLARGE,L1,XSMALL,IN,CIR,RADIUS,.5

9. By a radius and tangency to two circles (Fig. 137).

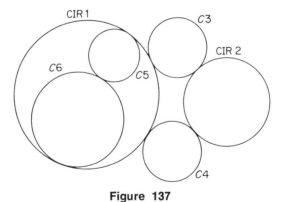

Figure 137

C3 = CIRCLE/YLARGE,OUT,CIR1,OUT,CIR2,RADIUS,.5
C4 = CIRCLE/YSMALL,OUT,CIR1,OUT,CIR2,RADIUS,.5
C5 = CIRCLE/XLARGE,IN,CIR1,OUT,C6,RADIUS,.5

11.6. CYLNDR DEFINITIONS

Cylinders normal to the XY plane, that is, with the axis of the cylinder parallel to the Z axis, may often be defined as circles. The cylinder, of course, extends indefinitely along its axis.

The definition of a cylinder requires that the axis of the cylinder be first determined as a vector. Usually a unit vector is defined. See Fig. 138 for typical unit vectors.

VEC1 is parallel to the Z axis. Its X, Y, Z components are, therefore, 0, 0, 1. VEC2 is parallel to the Y axis; therefore, its components are 0, 1, 0. VEC3 is at 45 deg to both X and Z axes and lies in the XZ plane. Its X, Y, Z components are .7071, .7071, 0. Vectors not passing through the origin may be defined by their end points; thus, VEC = VECTOR/P1, P2.

Suppose VEC3 represents the axis of a cylinder of radius 0.5. PTA is a point on the cylinder axis. The cylinder is thus defined

CYL = CYLNDR/PTA,.707,.707,0,.5 (Fig. 139)

CYL may also be defined by the coordinates of the point and the symbolic name of the vector:

CYL = CYLNDR/1,1,0,VEC3,.5

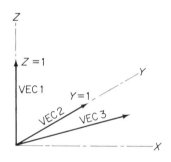

Figure 138

or again CYL = CYLNDR/PTA,VEC3,.5
or also CYL = CYLNDR/1,1,0,.707,.707,0,.5

giving the three coordinates of PTA, the three vector components of a unit vector, and the radius, in that order.

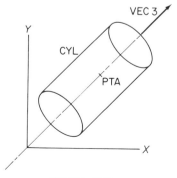

Figure 139

11.7. A PROGRAMMING SUGGESTION

LINE, PLANE, CIRCLE, and CYLNDR definitions will occur in most APT programs. Definitions of mathematically more complex surfaces must also be discussed, but to program the cutter over these more complex shapes still requires the assistance of the simpler surfaces.

Many programmers begin their manuscript of geometric statements with the following three definitions:

$$PT\emptyset = P\emptyset INT/0,0,0$$
$$LINX = LINE/PT\emptyset, ATANGL, 0$$
$$LINY = LINE/PT\emptyset, ATANGL, 90$$

This is a useful point and pair of lines for building up other definitions.

APT Practice Questions

1. Write APT definitions for L1, L2, and L3 of Fig. 11-1.
2. Write two possible APT definitions for each of the lines L1, L2, L3, and L4 of Fig. 11-2.
3. Write 5 possible definitions of C2 in Fig. 11-2.
4. In Fig. 11-3, there are 8 possible circles with the same radius as C1 and tangent both to C100 and L100. Write the definitions of all these possible circles.
5. See the O-ring groove, drawing no. 5 of part print no. 1. Write APT definitions of CIRC2 and CIRC4 based on tangency to CIRC3 and CIRC4. Such definitions actually would be preferable to those used in the sample APT program of Fig. 116. The definitions of Fig. 116 are accurate only to three decimal places, and the four circles are not necessarily tangent in terms of greater precision than 0.001".
6. Write a set of APT statements to define the contour of Fig. 11-4.
7. Write a set of APT definitions to define the contour of the part of Fig. 102. Omit the contours of the 0.187 plunge cut.

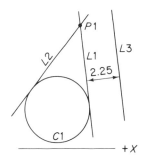

Figure 11-1

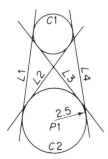

Figure 11-2

A Programming Suggestion   227

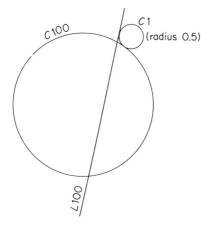

Figure 11-3

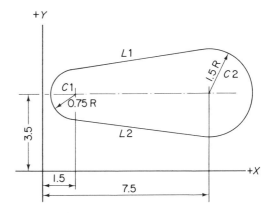

Figure 11-4

12

CONTROLLING THE TOOL MOVEMENTS

12.1. FIRST PRINCIPLES

The writing of the APT geometric statements is usually a more or less routine matter. Setting up the tool motion commands presents a number of minor difficulties and uncertainties for the learner of APT language, who may find that the computer takes a callous dislike to some of his first programs. There are, however, no serious difficulties to be expected, at least if his programming is confined to no more than three axes.

The learner should try first a few very simple programs using only a few simple tool commands, in order to establish the basic routines. After he thus has some concept of the caprices of his computer, he can become more ambitious in his undertakings. This chapter sets out only the most commonly met programming situations.

Tool motion commands are action verbs followed by a slash. Examples are GØTØ/, GØLFT/, GØFWD/. Specific details follow the slash. In programming the path of the tool, the assumption is made that the tool moves over the work piece while the latter is stationary. Usually, of course, the n/c machine moves the work piece; nevertheless, the programmer assumes the opposite in the programming.

The point on the tool that is programmed is the *tool end,* the intersection of the axis of the tool with the plane of the face of the tool, as shown in Fig. 140. The computer calculates the required tool offset to machine the part from the information in the CUTTER/ statement. The simplest cutter statement, and the usual one, is

Figure 140

CUTTER/(diameter).

Thus CUTTER/.5 means an end mill 0.500 in. in diameter. For n/c drafting, the cutter statement for a pen or pencil must necessarily be CUTTER/0. In the O-ring groove example, since there is no tool offset and the tool follows the contour, any cutter diameter, including CUTTER/0, could be specified.

In any n/c programming, the work piece and the spindle must be initially coordinated in the XY plane. This is a manual setup operation, discussed in Chapter 3. When tool movement is to begin, the tool end must start to move from some initial or parking position. The location of the tool end at the instant when the tool motions come under automatic control and machining begins is given by the point specified in a FRØM/ statement. See the O-ring program. The FRØM/ statement has two possible formats:

or
                      FRØM/X, Y, Z
                      FRØM/(point).

Thus FRØM/2,3,2 or FRØM/PT4. The point of initial programming is often named SETPT. At the end of the tool movement sequence the tool is returned to the point named in the FRØM/ statement.

For purposes of postprocessing, the feed rate in ipm may be added to a FRØM/ statement; thus

                    FRØM/2,3,2,20

12.2. GØTØ/ AND GØDLTA/

GØTØ/ is a positioning control statement. It directs the tool to go to the coordinates given in the statement. Thus

                    GØTØ/2,3,2

is an instruction to go to the point $X = 2$, $Y = 3$, $Z = 2$. Alternatively a symbolic name may be used:

                    GØTØ/SETPT.

The feed rate ipm may be added:

GØTØ/2,3,2,40

GØDLTA/ is an incremental statement, giving increments in X, Y, Z that the tool must move. Thus GØDLTA/3,0,−8 is a command to go from the present position +3 in. in X, no movement in Y, and −8 in. in Z.

Both GØTØ and GØDLTA may be clarified by drilling a hole with APT commands. Here is the program, which the learner should sketch if he requires additional orientation:

```
SETPT = PØINT/3,10,8
DRLPNT = PØINT/10,12,8
REMARK   HOLE TO BE DRILLED THRU AT 10,12
FRØM/SETPT
GØTØ/DRLPNT
GØDLTA/0,0,−3    $$DRILL FEEDS 3 INCHES INTO WORK
GØDLTA/0,0,3     $$DRILL RAPID RETRACT
GØTØ/SETPT
FINI
```

The drill, therefore, being at $Z = 8$, drills down −3 in. to $Z = 5$, retracts 3 in. back to $Z = 8$.

This little program is written for a three-axis machine. A two-axis machine such as the Cintimatic would require a slightly different program, since it uses a "canned" drilling cycle G81. This canned cycle would be called up by a postprocessor statement. But such statements are not the subject of this chapter. For the canned drilling cycle the program would read

```
FRØM/SETPT
GØTØ/DRLPNT
(postprocessor statement calling up G and M functions)
GØTØ/SETPT.
```

PRACTICE PROBLEM

Write an APT program for a three-axis drill to drill the four bolt holes in gear housing no. 2 of the gear pump.

12.3. ZSURF/

FRØM/ and GØTØ/ statements, and PØINT statements used in positioning operations are allowed to have only X and Y coordinates if so desired.

The Z coordinate is then taken to be zero, unless a ZSURF/ statement is used.

The ZSURF must be a plane; thus,

ZSURF/PLANE4

If then the following statements, for example, are used:

FRØM/3,6
GØTØ/5,12

the Z coordinate for these two statements is the Z coordinate that lies on the ZSURF at the respective XY locations.

The ZSURF need not be parallel to the XY plane, but may be tilted. If a number of holes must be drilled into a sloping surface, the calculation of Z coordinates to bring the tool clear of the part at each hole location would be tedious. Use of the ZSURF statement delegates such computations to the computer. The ZSURF statement must, of course, precede the point definitions and positioning commands; thus,

ZSURF/PLANE1
GØTØ/3,8
GØTØ/8,18

The Z coordinates for the positions 3, 8 and 8, 18 will be located on PLANE1.

12.4. TOOL MOVEMENT CONTROL IN CONTOURING

The method of controlling the movement of the tool by intersecting surfaces as discussed in this section is the basic principle of APT programming. The tool movement is controlled by three surfaces:

A drive surface
A part surface
A check surface

always specified in that order.

Consider a flat-end end mill cutting any contour, straight or curved. The end surface of the mill rides along one surface. The side of the cutter rides (and cuts) along another surface, in this case at right angles to the end surface. These two surfaces completely define the tool movement in space. For the case of the O-ring groove, the end surface (bottom plane of the groove) is the *part surface*. Usually this end surface is the part surface, though not in every possible case. The side controlling surface is the *drive surface*. This is the surface that guides the tool in producing the part shape.

These two surfaces, drive surface and part surface, define the path of tool motion in three dimensions, but do not provide initial or terminal points for that motion. These are provided by a third surface, the *check surface*. The check surface is some surface that intersects the tool path, to which the cutter must machine.

The computer calculates the tool positions with respect to these surfaces within the tolerance specified in the tolerance statements INTØL/ and ØUTTØL/. If no tolerance is specified, the APT routine assumes INTØL/0 and ØUTTØL/.0005.

The use of the three controlling surfaces is illustrated by Fig. 141.

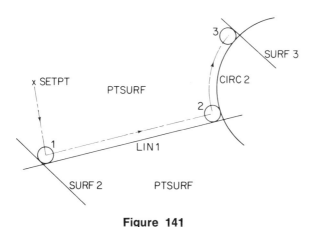

Figure 141

The tool has been directed to move off SETPT to LIN1 (recall that a line is considered as a plane). It must then follow LIN1 until it touches CIRC2, then go around CIRC2 to the check surface SURF3. The part surface is the plane of the page. To move the tool from position 1 to position 2, the tool motion command is

GØLFT/LIN1,ØN,PTSURF,TØ,CIRC2

GØLFT means "turn left." Left or right directions are the same as for driving an automobile, that is, the direction is as determined when facing in the direction of motion. The cutter is to turn left, following LIN1 as drive surface, with the tool end point ØN the part surface PTSURF. The tool is to follow LIN1 and to stop when just touching (TØ) the check surface CIRC2.

Note the order in which the three controlling surfaces must be given:

Drive surface
Part surface
Check surface

This order cannot be modified. The computer takes the first surface mentioned to be the drive surface, and likewise recognizes the other control surfaces according to their position.

To move the tool to position 3 along the required tool path, the command is

GØLFT/CIRC2,ØN,PTSURF,TØ,SURF3

Note that the check surface for the first move has become the drive surface for the second move. Again, this is usual.

The tool motion statement commences with an action verb followed by a slash. The three control surfaces are stated in their required order, and the relationship of the tool to the surface is defined by additional words such as TØ or ØN. Commas separate the words. No other punctuation other than the slash and commas may be used. Do not place a period at the end of the command.

12.5. THE START-UP STATEMENT

Two statements are required to initiate the sequence of cutter movements. One, the FRØM/ statement, specifies the initial or parked position of the tool (FRØM/SETPT). One more statement, the actual start-up statement, commands the first tool movement. This is a GØ statement, which will initially position the tool with respect to the first drive, part, and check surfaces.

The SETPT or starting point must not be located on or inside the contour of the part to be cut. To move the tool FRØM/SETPT to point 1 in Fig. 141 the start-up statement is

GØ/TØ,LIN1,ØN,PTSURF,TØ,SURF2

where LIN1 is the drive surface, PTSURF is the part surface, and SURF2 is the check surface. The succeeding motion statements to move the tool to positions 2 and 3 were given above. The complete set of statements to move the tool from SETPT to points 1, 2, and 3 are, then, these:

FRØM/SETPT
GØ/TØ,LIN1,ØN,PTSURF,TØ,SURF2
GØLFT/LIN1,ØN,PTSURF,TØ,CIRC2
GØLFT/CIRC2,ØN,PTSURF,TØ,SURF3

The part surface is defined in the GØ/ start-up statement, and does not change; therefore, it may be omitted from the following two GØLFT/ statements as follows:

GØLFT/LIN1,TØ,CIRC2
GØLFT/CIRC2,TØ,SURF3

The difference between a GØTØ/ and a GØ/TØ, statement must be clearly understood. If the use of these two statements is interchanged, the computer will be unable to process the program. GØTØ/ is a positioning statement: "Go to a point." GØ/TØ, means "Go to a set of control surfaces."

The GØ/ statement requires the use of the modifiers TØ, ØN, PAST, and TANTØ. These are illustrated in Fig. 142. The tool, there-

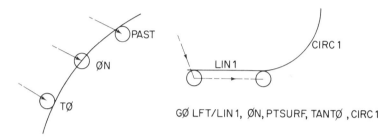

GØ LFT/LIN 1, ØN, PTSURF, TANTØ , CIRC 1

Figure 142

fore, may

GØ/TØ (just touching the surface on the near side)
GØ/ØN (with the tool end on, or within tolerance of, the surface)
GØ/PAST (just touching the surface on the far side)

A fourth very common case arises, where the check surface is tangent to the drive surface (TANTØ). This is also shown in the figure. The TANTØ modifier, however, may be used only with respect to a check surface.

If no modifier, TØ, ØN, PAST, or TANTØ, is used, then the APT computer system assumes the use of the modifier TØ; thus

GØ/DS,PS,CS

means GØ/TØ all three surfaces.

A feed rate may be added to a GØ/TØ statement:

GØ TØ,DS,ØN,PS,PAST,CS,6.

12.6. GØ MOVEMENT STATEMENTS

In directing the tool to follow a contour, six possible direction commands may be employed:

GØLFT
GØRGT

GØFWD
GØBACK
GØUP
GØDØWN

The commands GØFWD or GØBACK are used at points of tangency, as illustrated in Fig. 143. GØLFT and GØRGT are applicable over the wide angle shown in Fig. 144. Since GØFWD may also be applied over 176 deg, the angular region over which either GØFWD or GØLFT (or GØRGT) would equally well apply is 86 deg. However, the command that more reasonably applies is selected.

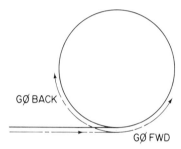

Figure 143

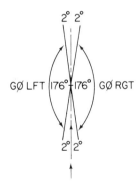

Figure 144

GØUP and GØDØWN are not commonly required, and should be used when no other direction command is applicable. GØUP means that the tool is withdrawn, whereas GØDØWN indicates a plunge cut toward the work. If GØUP or GØDØWN should be used, the sense of direction for the following motion is taken as that of the last motion command before GØUP or GØDØWN, provided that UP and DOWN are within 2 deg. of the spindle axis.

12.7. SPECIAL CASES IN TOOL MOTION

1. An Ambiguous Case

In Fig. 145, if the tool command is GØ/S1,S2, then the computer

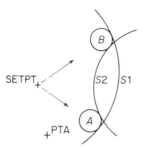

Figure 145

is offered two possibilities: The tool may go to position A or to position B. This ambiguity must be removed with the use of an additional statement, either INDIRP or INDIRV.

The command INDIRP means "in the approximate direction of the given point." PTA is shown in the figure. For the tool to go to position A, the following statements would be used:

FRØM/SETPT
INDIRP/PTA
GØ/TØ,S1,TØ,S2

Alternatively, INDIRV may be used. This statement means "in the general direction of the given vector." The vector is defined by giving its X, Y, and Z components. Either INDIRP or INDIRV may be used as required anywhere in the tool movement program. Although a vector pointed in the general direction desired is sufficient, it should not be more than the cutter diameter from the actual direction. The vector also should pierce both the drive and check surfaces of the GØ/ statement.

FRØM/SETPT
VEC = VECTØR/1,1,0
INDIRV/VEC
GØ/TØ, etc.

Either INDIRV or INDIRP may be used with a two-surface start-up statement, the check surface being omitted. The tool will then adopt the general direction given by INDIRV or INDIRP when coming into tolerance of the given drive and part surface.

2. PSIS/

Sometimes the part surface established by the start-up statement can no longer be used. Perhaps the drive surface or check surface of a motion statement must become the new part surface. This new part surface must be either the drive or check surface of the immediately preceding motion command, because the cutter must already be within tolerance of the newly designated part surface. This new part surface is specified in a statement thus:

PSIS/

PSIS/PL1

or other symbolic name.
The new part surface remains in effect until superseded by another part surface.

3. TLLFT, TLRGT

These words specify that the tool shall be on the left side or right side of the drive surface, tangent to that surface, when looking in the direction of motion; thus,

	TLLFT,GØLFT/DS, etc.
or	TLLFT
	GØLFT/DS, etc.

These words should be included in the motion commands only where required. Sometimes, when the computer cannot seem to process the program, a TLRGT (TLLFT) helps.

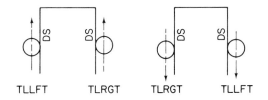

Figure 146

4. TLØNPS, TLØFPS

These modifiers remain in effect until changed by another command; that is, like PSIS, or TLLFT, they are "modal" or continuing. They mean "tool on part surface" and "tool off part surface." As indicated in Fig. 147, TLØNPS indicates that the tool end is to follow the part surface, with the trailing or leading part of the tool gouging the surface. TLØFPS

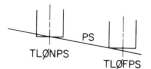

Figure 147

means that only one point of the tool touches the part surface, which is not gouged by the tool, therefore. Of course, if the tool end surface and the part surface are parallel, there is no need for these commands. They are employed thus:

> TLØNPS
> GØRGT/DS, etc.

5. Multiple Intersections

This is a common circumstance. In Fig. 148 the tool is to machine entirely around the circle, using S1 as check surface. Suppose the following sequence of commands is written:

> FRØM/SETPT
> GØ/PAST,S1,ON,PS,TO,C1
> GØLFT/C1,PAST,S1

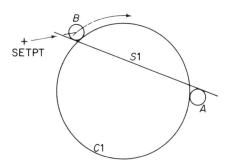

Figure 148

The tool will stop at position A, instead of circumscribing the circle to the point B. Of course, another statement GOFWD would continue the motion beyond A. But for this case the INTØF modifier is employed:

> FRØM/SETPT
> GØ/PAST,S1,ØN,PS,TØ,C1
> GØLFT/C1,PAST,2,INTØF,S1

The command calls for the tool to go PAST the second intersection (2,INTØF,) of the two surfaces, bringing it to the terminal point B.

6. THICK/

The THICK/ statement calls for calculated tool positions to be displaced by the amount in the THICK/ statement, resulting in an excess of metal on the part if a plus thickness is specified, or an undersize part if a minus thickness is called for. For example,

>FRØM/PT1
>THICK/.1,.1,0
>GØ/PAST,C1, etc.

The three thickness measurements apply as follows:

>The first one to the part surface (0.1 in.)
>The second one to the drive surface (0.1 in.)
>The third one to the check surface (0.0 in.)

If only one THICK distance is given, it applies to all three surfaces.

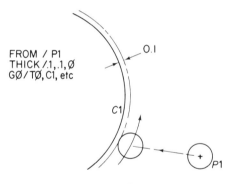

Figure 149

7. A Special Case

In Fig. 150 the two statements shown are written to bring the tool to the intersection of C1 and LIN1. Any such sequence should fail if it tries to bring the tool across C1 to get at the intersection of C1 and LIN1. Let us say that the computer does not have "line of sight" to LIN1 because the nearer part of C1 obscures the sight. To improve the configuration, move the tool to the right before writing the GØ/ start-up statement, as follows:

>FRØM/SETPT
>GØDLTA/2,0,0
>GØ/ØN,LIN1,ON,C1

12.8. PROGRAM PRINTOUTS

1. The computer printout will always include the manuscript program.

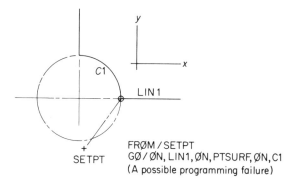

Figure 150

2. The printout of the successive tool path locations is called up by the command CLPRNT. This command may appear anywhere in the program.
3. Additional information may be wanted for geometric definitions. The programmer, for example, might wish to know the coordinates of the center of a circular fillet. This additional information can be printed out by using the command PRINT/3,ALL. This will produce a printout of all information in computer memory for all geometric definitions preceding the PRINT/3,ALL command.
4. The PRINT/3,ALL printout will immediately follow the CLPRNT on the page. If the PRINT/3,ALL information is to be printed at the top of a new page, add the additional statement PRINT/0 as follows:

> CLPRNT
> PRINT/0
> PRINT/3,ALL

12.9. SOME OBSERVATIONS AND SUGGESTIONS FOR PROGRAMMING IN APT

1. The start-up GØ/ statement does not really have a drive surface. It serves the following purposes:

a. It brings the cutter to the contour.
b. It defines the part surface for the initial machining movement and subsequent movements.
c. It sets up the drive surface for the initial machining movement.
2. The part surface is defined in the start-up statement and need not be included in the following motion statements. The part surface thus defined is, therefore, "modal," as previously explained. Any modal word continues in effect throughout the program until countermanded.
3. The contour is preferably followed by the cutter in the counter-clockwise direction.
4. The tolerance may be changed at any point in a program by the use of the TØLER/ statement. This specifies a plus tolerance and an assumed (but not specified) zero minus tolerance, thus: TOLER/.00015. However, if the TØLER/ statement reduces the tolerance, a new start-up statement will probably be necessary, since with the tighter tolerance the tool will probably be put out of tolerance of the existing control surfaces.

12.10. NESTED DEFINITIONS

If a geometrical item is needed only once in a program and will not be used again, it may be used where needed in another geometrical definition by means of a nest in parentheses. An example will explain such usage. Consider the following case, where two points are defined in order to define a line through these points.

$$PT1 = PØINT/1,2,3$$
$$PT2 = PØINT/4,5,6$$
$$LN1 = LINE/PT1,PT2$$

Now suppose that there is no further use in the program for PT1 and PT2. The amount of programming may be reduced by nesting these point definitions in the line definition, thus reducing three program statements to a single statement:

$$LN1 = LINE/(PT1 = PØINT/1,2,3),(PT2 = PØINT/4,5,6)$$

Parentheses are required, with a comma between the two nested definitions.

The symbolic name need not be included in the nested definition; thus:

$$LN1 = LINE/(PØINT/1,2,3),(PØINT/4,5,6)$$

Questions

1. Program the O-ring groove of part print no. 1 for an n/c drafting machine. Use the command DRAFT/ØN to commence drawing a line and DRAFT/ØFF at the end of a line to be drawn. Number your statements, 5, 10, 15, 20, etc.
2. Using APT positioning commands, program the drilling of the two dowel holes in gear housing 1 of part print no. 2.
3. Sketch the part produced by the following program and indicate with small circles the positions of the end mill on the tool path.

```
PARTNØ   LINEAR CAM
CUTTER/.25
SETPT = PØINT/−3,−5
LIN1  = LINE/0,0,6,0
LIN2  = LINE/9,0,9,5
LIN3  = LINE/0,0,6,3
LIN4  = LINE/PARLEL,LIN1,YLARGE,3
LIN5  = LINE/PARLEL,LIN2,XSMALL,9
FRØM/SETPT
GØ/TØ,LIN5,TØ,LIN1
GØRGT/LIN1,PAST,LIN2
GØLFT/LIN2,PAST,LIN3
GØLFT/LIN3,TØ,LIN4
GØRGT/LIN4,PAST,LIN5
GØLFT/LIN5,PAST,LIN1
GØTØ/SETPT
FINI
```

4. Using a SETPT at the middle of the pocket, write an APT program to mill around the periphery of the large pocket of part print no. 2. Start at the upper left-hand corner of the pocket and mill counterclockwise. Omit any movements in Z. Omit also the part surface, using only drive and check surfaces.
5. Program the part shown in Fig. 12-1, including the following statements at appropriate places in the program:

```
TOLER/.003 (instead of INTØL and ØUTTØL)
SPINDL/ØN,400
SPINDL/ØFF
```

6. Figure 12-2 is the drawing of a punch plate used to carry and support the punch in punch press operations. Such plates are flame-cut.

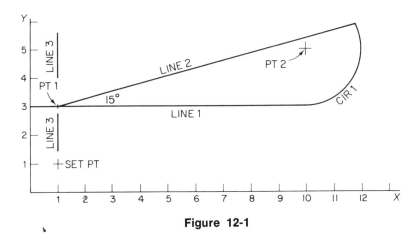

Figure 12-1

Program the cutting of this plate for an n/c flame-cutting machine. The SETPT is to be located at X = 4, Y = 13. The cutting oxygen must be on before the cutting head moves off SETPT, since SETPT is at the edge of the plate. Use the commands FLAME/ØN and FLAME/ØFF for control of the cutting oxygen (these commands are not standard for all flame-cutting machines).

For part surface, use the name PARPLN. Begin the contouring at the intersection of L1 and C1 and cut clockwise around the plate periphery. The following statements also are required:

CUTTER/.19
INTØL/.003
ØUTTØL/.006
MACHIN/FIRBR,1 (a postprocessor for a fictitious machine)

7. The part shown in Fig. 12-3 is an aluminum bracket ½ in. thick. The XY plane is to be taken as the part surface (PARPLN) and is 0.600 in. below the top surface of the bracket as affixed to the machine. No clamps interfere with the machining of this contour. The geometric elements, lines and circles, can be defined as lying in the XY plane for convenience, since a LINE is a plane and a CIRCLE is a cylinder. Sufficient information is supplied in the figure to define all geometric elements of the contour.

The contour should be machined counterclockwise. The tool, a 0.750-in. end mill, must GØDLTA/ in Z at the start and at the finish of contouring. Commence machining at least 0.1 in. off the finished contour, which is rough-sawed.

8. The leaf cam of Fig. 12-4 is a flat plate of 1040 steel ¼ in. thick. It is affixed so that the top surface is ¾ in. above the work table. No clamps are used on the periphery of the contour. In the fully re-

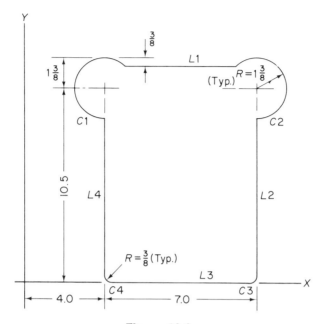

Figure 12-2

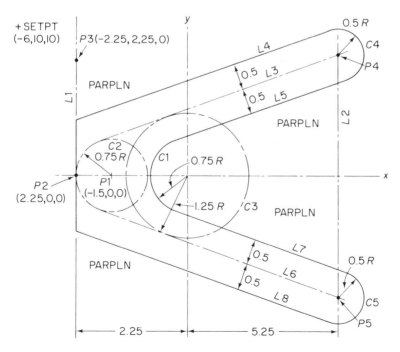

Figure 12-3

tracted position, the tool end is 6 in. above the work table. Program the part in APT for a ¾-in. end mill.

Include the following postprocessor commands in the program:

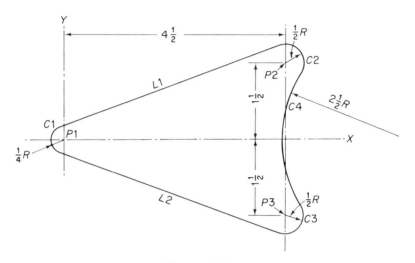

Figure 12-4

MACHIN/MILMAR,3 (a fictitious machine)
SPINDL/ØN,440,CLW
SPINDLE/ØFF
CYCLE/MILL (before and after the milling operation, to call up the required G function)
ØRIGIN/3,10,0 (to translate the origin of coordinates as set up on the work table 3 in. in X and 10 in. in Y from the origin used in the APT program)

9. The work piece of Fig. 12-5 is delivered to an n/c vertical mill as a flat piece of 2024 aluminum 1 in. thick, with the finished triangular shape P1P8P3. It requires an APT program for milling out the ledge around the periphery with a flat 1.000 end mill.

Define all drive, part, and check surfaces as planes and program the ledge cuts. Is a TLØNPS or TLØFPS instruction needed?

10. Program the contour of Fig. 12-6 as a two-dimensional contour, omitting any Z axis considerations. Do not program the holes.

11. If the O-ring groove program of Figs. 116 and 117 is given to the computer with successively smaller tolerances, .002, .001, .0005, .0002, etc., a tolerance will be reached that is small enough to cause the program to fail. The failure will occur at the GØFWD statements from one CIRC to the next. Can you account for the failure? If you can, how must you change the program to correct this failure?

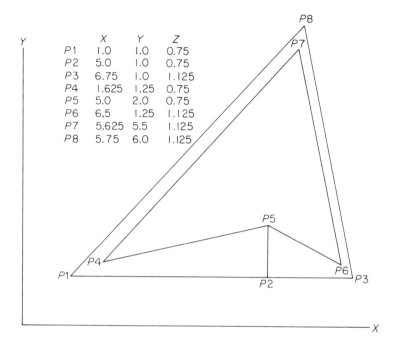

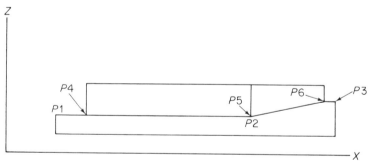

Figure 12-5

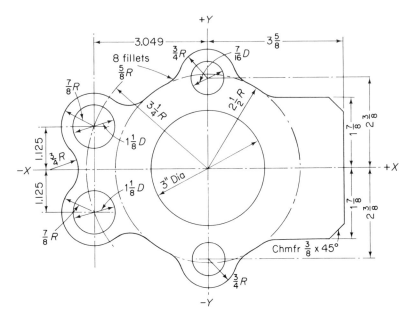

Figure 12-6

13

THE POSTPROCESSOR

13.1. GENERAL CONCEPTS

The APT system produces a general mathematical solution to the cutter path sequence in the main processor. The adaptation of this generalized solution to the requirements of the specific machine to be used is executed by reprocessing the computer output in a second pass through the computer, the postprocessor.

The requirement of two computer programs in order to obtain data for a punched tape roughly doubles the computer cost and perhaps increases the programming cost. It is possible to invent n/c computer methods that complete the job in one computer input, and indeed, a considerable number of such programs are available: PATH DECK (Monarch Machine Tool Company, for their n/c lathes), SPLIT (Sundstrand Corporation, for multi-axis mills), SNAP (Brown & Sharpe, for drills), and others. Usually, such programs are suited to one type or one brand of n/c machine only. The advantage of the postprocessing method may be illustrated by assuming that 20 of a certain part must be produced on a Kearney & Trecker Milwaukee-Matic Model II, with another 20 of the same units to be subcontracted to a factory using a Gorton Tapemaster 2-30, while at the same time the part must be drafted on an Orthomat n/c drafting machine. The same general solution serves all three machines, since it is a mathematical solution to the problem unmodified by hardware considerations. The main processor solution must be postprocessed three times to meet the requirements of the three machines and their machine control units. The required postprocessor programs are called up by these three statements:

1. MACHIN/BXKT23   Kearney & Trecker Model II equipped with Bendix Dynapath machine control unit
2. MACHIN/BR3100,7 Gorton Tapemaster 2-30, three axis, with photoelectric tape reader, Bunker-Ramo control
3. MACHIN/TRW,2,1   Orthomat equipped with Thompson-Ramo-Woolridge machine control unit

Note that the identifier name that calls up the required postprocessor routine identifies both the controller (Bunker-Ramo 3100) and the machine tool (7).

The postprocessor program has many functions:

1. It must round off the computer CLPRNT figures, often eight, ten, or more digits in length, to the number of digits required for the tape format, which will be five or six. Leading and trailing zeros may have to be added. The decimal point must be removed.
2. It must supply the required tape format, word address or tab sequential, and EIA or ASCII character coding as required.
3. It may have to change the origin of coordinates as used in the APT program to a new origin for machine setup.
4. It must provide printout information for diagnosis and checking.
5. It must check that the tool movements called for in the main processor do not exceed the size and movement limitations of the machine tool.
6. It must control acceleration and deceleration to prevent overshoot of the work table.
7. It determines and prints out the length of the n/c tape.

The postprocessor commands are interspersed appropriately throughout the program manuscript. The command NØPØST (no postprocessing) or MACHIN/ are both placed near the beginning of the program. Both are postprocessing commands. The last word in the following statement is a feed rate number in ipm, and likewise is a postprocessor command:

GØ/DS,ØN,PS,TØ,CS,6

A CØØLNT/ØN command, also a postprocessor statement, naturally must precede the tool motion statements; CØØLNT/ØFF must follow the set of tool motion statements.

Because every n/c machine-controller combination has its individual requirements, it is possible here only to make some general comments on postprocessing. It would be neither useful nor proper to demonstrate at length the postprocessing procedures for one particular machine. The programmer must have the postprocessing manual applicable to his own n/c machine. A few postprocessing attempts will make him familiar with the procedures. The following sections refer to the more commonly used postprocessor commands, concluding with some statements used in n/c flame-cutting and drafting.

13.2. THE MACHIN/ STATEMENT

As previously explained, this statement calls up the set of instructions (postprocessor program) that is required to convert the APT general solution to the form required for the machine tool. Certain computers and computer systems have proprietary postprocessing methods and may use different formats for calling up the postprocessor. Some MACHIN/ statements for more familiar n/c machines are the following:

Cincinnati vertical spindle two-axis mill	CINMAT,1
Cincinnati horizontal spindle two-axis mill	CINMAT,2
Gorton 2-30 three-axis contouring mill, 100 cps reader	BR3100,6
Kearney & Trecker Milwaukee-Matic Model II, contouring	BXKT23
Kearney & Trecker Milwaukee-Matic Model II, four-axis	BXKT24
Compudyne Model B, three-axis mill	CMPDYN,1
Compudyne Model C, three-axis mill	CMPDYN,2

13.3. SPINDL/

The SPINDL/ statement causes the postprocessor to produce the correct codes for the spindle command.

SPINDL/660,CLW indicates 660 rpm, clockwise.
SPINDL/395,CCLW indicates 395 rpm, counterclockwise.
SPINDL/ØFF generates an M05 code (spindle off).

The SPINDL/ command is modal; that is, it is maintained until countermanded.

These postprocessor SPINDL/commands apply only to machines with spindle operation under tape control.

13.4. CØØLNT/

CØØLNT/ØN will provide flood coolant if no coolant has previously been specified. In some postprocessors this command generates an M08 code (coolant on).

> CØØLNT/ØFF generates an M09 function (coolant off)
> CØØLNT/FLØØD
> CØØLNT/MIST

13.5. FEDRAT/

FEDRAT/12 will generate an F code of 12 ipm along the path of movement.

Feed rates may also be attached to motion statements, as explained in the previous chapter. A feed rate is modal (remaining in effect until cancelled).

13.6. TRANS/ OR ØRIGIN/

This statement is usually programmed before the FRØM/ start-up statement.

The programmer has probably selected a Cartesian coordinate system for programming the part which offers the most computational convenience, and without regard to the coordinate system of the machine tool. He will use the postprocessor for converting the part coordinate system used into the system to be used when machining, by the employment of the TRANS/X, Y, Z statement. Refer to Fig. 151 for explanation.

The part program coordinate system is represented by XYZ and

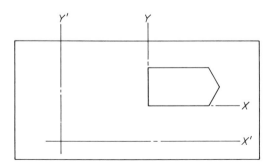

Figure 151

the machine-tool coordinate system by X'Y'Z'. Suppose that the origin of the machine system is at $X = -10$, $Y = -3$, $Z = 0$ with respect to the part coordinate system. Then the following statement transforms the part system to the machine system:

$$\text{TRANS}/10,3,0$$

The computer will then add 10 in. to the programmer's X dimensions and 3 in. to his Y dimensions. Some postprocessors are ØRIGIN/ instead of TRANS/.

13.7. SEQNØ/

This command is used to insert sequence numbers into the information blocks of the tape. The format of this command may vary somewhat among postprocessors.

> SEQNØ/ØN,INCR,5 indicates that sequence numbering should begin and be incremented by 5 each time.
> SEQNØ/100,INCR,1 commands sequence numbering to begin, starting with 100 and incrementing 1 each time.
> SEQNØ/ØFF terminates sequence numbering.

13.8. MCHTØL/

MCHTØL/ means machine tolerance. The command controls overshoot when the movement of the machine changes direction, as in Fig. 118. The machine feed rate may have to be decelerated to prevent the overshoot shown. MCHTØL/.002 is a statement requiring that machine inertia effects be controlled to limit overshoot to 0.002 in.

13.9. STØP

The STØP command generates the miscellaneous function M00; that is, all slide motion, coolant, spindle, and tape reader stop, but the machine control unit is not turned off. The operator pushes the proper start button to continue the n/c tape-controlled operation.

13.10. END

This statement shuts down the n/c machine, including the machine control unit, at the end of the program by generating an M02 function. It also produces a trailer on the end of the tape for ease of tape feeding. This command usually appears in the program immediately before the main processor command FINI.

13.11. LEADER/

The LEADER/ statement generates a length of tape leader as specified in the statement. Thus LEADER/18 produces 18 in. of leader at the front of the tape.

13.12. AUXFUN/

This statement calls for a miscellaneous function or for turning it off. Thus AUXFUN/7 in some postprocessors will generate an M07, while in others AUXFUN/07 does the same thing. The auxiliary function is inserted into the next block of information. Also used:

> AUXFUN/ØN
> AUXFUN/ØFF

13.13. CLEARP/ AND RETRCT

CLEARP/ specifies a clearance plane to which the tool is withdrawn when a RETRCT command is issued. The following is an example for a three-axis mill or drill:

> CLEARP/XYPLAN,8
> GØTØ/3,2,1
> RETRCT

The CLEARP/ statement defines the clearance plane as a plane parallel to the XY plane, with $Z = 8$. This plane is selected far enough back to prevent any collision due to rotation of a work table, or for any other reason. The tool goes to 3,2,1, then retracts back in Z to CLEARP.

13.14. CYCLE/

This statement sets up a machining operation to be executed at the points following the command until the CYCLE/ØFF statement is given. This postprocessor may in some systems generate a G function such as G81 (drill). The command may be any of the following or others, though for some n/c machines the CYCLE/ command may have many words for a more complex machine movement:

    CYCLE/AVOID (retract to avoid a clamp, etc.)
    CYCLE/BØRE
    CYCLE/DEEP (deep hole drilling)
    CYCLE/DRILL
    CYCLE/MILL
    CYCLE/REAM
    CYCLE/ØN
    CYCLE/ØFF

13.15. PREFUN/

This command calls for the preparatory function specified, thus PREFUN/1 (G01, linear interpolation for a lathe).

13.16. DELAY/T

This statement specifies the number of seconds of dwell, thus DELAY/2.

13.17. LØADTL/

The command calls for an automatic or manual tool change to the tool number specified after the slash.

13.18. RAPID

The command RAPID causes the next motion to be executed at rapid feed rate. Feed rate then returns to the previously programmed rate for the following commands.

13.19. SELCTL/

This command selects the tool to be used next, putting it in position for automatic tool changing when the next LØADTL/ command is read.

13.20. RØTABL/

A command to rotate the table clockwise to an absolute angular position, thus:

    RØTABL/ATANGL,0

13.21. PLUNGE/

This statement produces motion parallel to the spindle axis for drilling or tapping. The distance of the plunge and the ipm are specified thus:

PLUNGE/.875,1.5

13.22. FLAME/

In automatic flame-cutting, the FLAME/ statement calls up various conditions:

FLAME/PREHET,ØN
FLAME/ØXYGEN,ØN
FLAME/ØXYGEN,ØFF
FLAME/ MASTER, ØXYGEN,ØFF
SLAVE,

13.23. DRAFT/

The DRAFT/ command controls pen operations and line types in automatic drafting.

DRAFT/ØN   DRAFT/DØTTED
DRAFT/ØFF   DRAFT/SØLID
                  DRAFT/DASH
                  DRAFT/DITTØ
                  DRAFT/CTRLIN

The command may include type of line, color, and intensity of line, as in the following examples:

DRAFT/DOTTED,PEN,BLACK,INTENS,LITE
DRAFT/CTRLIN,PEN,RED,INTENS,MEDIUM
DRAFT/DASH,PEN,BLUE,INTENS,DARK

If no specifications are given, simply DRAFT/ØN, then a solid line of medium intensity is implied.

Other drafting commands:

PENDWN   positions the pen on the drafting table
PENUP   positions the pen above the table

The place of a DRAFT/ command in the APT tool motion sequence follows the dictates of common sense. The tool movement FRØM/SETPT to contour is not drawn. The statement sequence would be

>FRØM/SETPT
>GØ/ØN,DS,ØN,PS,ØN,CS
>DRAFT/SOLID,PEN,BLACK,INTENS,MEDIUM
>DRAFT/ØN
>GØFWD/ etc.

The drafting stylus does not go TØ or PAST, but ØN or TANTØ. DRAFT/ØFF should precede the final tool movement GØTØ/SETPT.

13.24. POSTPROCESSING WITHOUT A POSTPROCESSOR MANUAL

Cases have arisen where APT programs had to be postprocessed without the help of a postprocessing manual. Such attempts probably are good experience. It is possible to make good guesses for the postprocessor commands for such machines as the Cintimatic vertical spindle mill or the Milwaukee-Matics, but perhaps not so easy for the Compudyne Contoura.

If the reader wishes to try his skill in this guessing game, he should attempt to postprocess the O-ring groove (part print no. 1) program for, say, the Cintimatic vertical spindle mill (MACHIN/CINMAT,1). He will probably require three attempts to get it right, which will mean a computer cost of about $30.

14

VERSATILITY IN APT

This chapter includes a number of special APT routines of considerable usefulness: the management of hole patterns in positioning operations, the convenient MACRO routine, and the determination of surface areas, centers of gravity, and moments of inertia of irregular shapes. The PØCKET routine may be extended to such problems as the determination of the best sequence of movements for combining a field of grain.

14.1. MOMENT OF INERTIA AND SURFACE AREA

Unfortunately, few computer services are able to handle the routines discussed in this section. The routines are mentioned, however, since they are of interest to colleges for the teaching of applied mechanics, strength of materials, and other subjects.

In design work and material cost estimating, such information as weights, surface areas, centroids, and moments of inertia is needed. The APT system includes routines for procuring such data.

Suppose that the following information is desired for the half-ellipse of the O-ring groove, part print no. 1:

1. Its length.
2. The area bounded by the curve of the half-ellipse.
3. The X and Y coordinates of the centroid of this area.
4. Moments of inertia of the area within the half-ellipse with respect to X and Y axes.

5. Moments of inertia with respect to axes parallel to the X and Y axes and passing through the centroid.

The APT program is written as for machining or drafting but using CUTTER /0 and one more modification, depending on the APT MACHIN/ statement used. There are four of these special machine statements.

MACHIN/COMPTR,INERTA,1. This statement is used when the X axis is the reference axis and the described curve lies on or above this axis.

Part of the APT tool motion sequence includes the path from the SETPT to the curve. This is not desired in these calculations. Other movements in the sequence may not be desired also. The tool movements that constitute the desired area are enclosed within two ITEM/1 statements thus:

        FRØM/SETPT
        GØ/ØN,LIN1,ØN,PTSURF,ØN,LIN2
        ITEM/1
           - - - - - -
           - - - - - -
           - - - - - -
        ITEM/ØFF
        GØTØ/SETPT
        FINI

MACHIN/COMPTR,INERTA,3. This statement is used when the Y axis is the reference axis and the described curve lies on or to the right of the Y axis. ITEM/ statements are used as required.

If the closed path of the curve is traced counterclockwise the sign of the calculated quantities will be positive; if clockwise, negative. If the part has holes or cutouts, they can be traced clockwise so that they are subtracted from the total area, which is traced counterclockwise.

Next suppose that the half-curve of the O-ring is to be rotated about an axis of revolution to produce a surface of revolution. Again the area is traced with a CUTTER/0 statement, and a MACHIN/ statement will provide information on this surface of revolution:

        MACHIN/CØMPTR,INERTA,2

This statement is used if the X axis is the axis of rotation and the generating curve lies on or above this axis.

If the Y axis is the axis of rotation and the generating curve lies on or to the right of the curve, the MACHIN/ statement is

        MACHIN/CØMPTR,INERTA,4

The ITEM/ statements apply as before. A DENSE/ statement may be inserted after the ITEM/1 statement. This will supply the density (specific weight) of the material in pounds per cubic inch, grams per cc, or other unit. If there is no DENSE/ statement, unit density of course results.

With either of these two MACHIN/ statements, the computer output provides the following data:

1. Length of the generating curve.
2. Weight of the solid of revolution produced by the rotation.
3. Surface area of this solid.
4. X and Y coordinates of the center of mass.
5. Moments of inertia with respect to X and Y axes.
6. Moments of inertia with respect to axes parallel to both X and Y axes and passing through the center of mass.

14.2. THE MACRØ FEATURE

Suppose that several holes must be drilled by means of APT statements in a program such as the following:

```
         GØTØ/6.5,4,1        $HØLE 1
         GØDLTA/0,0,—1
         GØDLTA/0,0,+1
         GØTØ/8,4,1
         GØDLTA/0,0,—1
         GØDLTAØ,0,+1
         GØTØ/9.5,4,1
         GØDLTA/0,0,—1
         GØDLTAØ,0,+1
         GØTØ/11,4,1
              etc.
```

There are three statements per hole, and if there are 12 holes, then 36 statements must be written and 36 cards punched. This is a great deal of programming effort for a small amount of information processing. But the same operation is repeated at each hole location, and by the use of a MACRØ statement as a subroutine for this operation, the number of statements may be reduced.

A MACRØ is a single statement that calls up a group of statements as needed in a program. In the program above, the operation of each position is a drill-retract sequence. Suppose that this sequence is named DRLRCT or some other symbolic name. The rules for MACRØ symbolic names are those for other symbolic names selected by the programmer. The repeated operation is defined by means of the MACRØ feature thus:

DRLRCT = MACRØ
GØDLTA/0,0,−1
GØDLTA/0,0,+1
TERMAC

These statements set up a packaged routine that can be called up in the program as required. The first statement identifies the routine by its symbolic name DRLRCT. The information or sequence of the routine follows. The MACRØ is terminated by the single word statement TERMAC.

The MACRØ sequence must be set up somewhere in the program before the first place where it will be called into execution. The MACRØ is executed by means of a CALL/(macro) statement thus:

DRLRCT = MACRØ
GØDLTA/0,0,−1
GØDLTA/0,0,+1
TERMAC
GØTØ/6.5,4,1
CALL/DRLRCT
GØTØ/8,4,1
CALL/DRLRCT
GØTØ/9.5,4,1
CALL/DRLRCT
     etc.

The MACRØ is processed statement by statement exactly as defined whenever called up in the program. The program for drilling 12 holes has been reduced from 36 statements to 28 by means of the MACRØ. But there is another repetitive operation here: the positioning of the spindle to the holes. The Y and Z values of the pattern of holes do not change, but the X values change. This hole pattern could be executed with the use of a PATERN/ statement, explained later in this chapter, or with a MACRØ.

If there is a variable in the MACRØ, it is included in the MACRØ definition after a slash. In the present case the MACRØ would be defined

MACRØ/X

indicating that X is a variable that will be defined each time the MACRØ is called for execution. The revised drilling program would be the following:

DRLRCT = MACRØ/X
GØTØ/X,4,1
GØDLTA/0,0,−1
GØDLTA/0,0,+1
TERMAC
FRØM/SETPT

```
            CALL/DRLRCT, X = 6.5
            CALL/DRLRCT, X = 8
            CALL/DRLRCT, X = 9.5
                 etc.
            GØTØ/SETPT
            FINI
```

This revised MACRØ reduces the program to a reasonable number of statements. (It should be apparent that APT is not a convenient language for positioning operations.)

The MACRØ is not restricted to a single variable.

As a second example of the use of the MACRØ, consider the employment of this programming device for rough and finish cuts, the finish cut to use the same tool path and a larger cutter diameter. The program will be such as the following:

```
            MACUT = MACRØ/DIA
            CUTTER/DIA
            FRØM/SETPT
            GØ/ØN,L1,ØN,PS,ØN,L2
            GØLFT/L2,ØN,L3
                 etc.
            GØTØ/SETPT
            TERMAC
            CALL/MACUT, DIA = 1.0
            LØADTL/MANUAL (manual tool change)
            CALL/MACUT, DIA = 1.1
            FINI
```

14.3. THE PØCKET/ ROUTINE

PØCKET/ routines are for pocket milling, but may be applied to face milling and slabbing-off operations, or even to nonmachining operations that must cover a defined area.

To clean out a pocket with an end mill requires sometimes a considerable number of geometrical statements. The PØCKET/ command can produce the whole sequence of motions in a single command. In this routine, the end mill drills into the approximate center of the pocket, then takes out the material in a series of increasingly larger polygons until the boundaries of the pocket are reached.

The pocket is defined by its corner points, and the order in which the points are given determines the path of the cutter. At the end of the pocketing operation the cutter must be brought out of the pocket vertically by another command, such as GØDLTA/, before further tool motions are given. The programmer does not know the point near the pocket center that the computer will choose to initiate the pocketing operation. Before the PØCKET command, the tool should be positioned in such a way that the cut toward the pocket center will do no harm; the movement to the bottom of the pocket is usually a ramp cut. The initial position of the tool, therefore, should be just inside the pocket. The example to follow will make this advice clearer.

The PØCKET routine cannot be used on pockets that contain an angle greater than 180 deg. Such a pocket would have to be divided into two pockets. The routine also does not allow for leaving islands of material in the pocket.

The PØCKET commands require considerable information:

PØCKET/Re,C,f,F1,F2,F3,0,P,PT1,PT2,PT3, . . . PTn.

1. Re. Effective radius of the cutter. For a flat end mill such as a slot drill, this effective radius could be the actual or full radius. If overlapping cuts are to be used, then Re is a radius less than the full radius. If a ball end mill is used, then Re must be selected to give a controlled height to the scallops left on the surface by the tool.
2. C. This is the cutter bite, or offset between parallel passes divided by the actual cutter radius (actual cutter radius, not Re). This number cannot be greater than 2, since the maximum offset between passes is evidently 2 cutter radii. If the bottom of the pocket is sloped, then C must be the offset in the plane of the pocket bottom.
3. f. This component is the depth of the finishing cut or the last cut along the sides of the pocket divided by the actual cutter radius. If the final cut is to be 0.060 and cutter radius is 0.500, then $f = .120$. If no special finish is desired, set $f = C$. As with C, f must be calculated in the plane of the bottom of the pocket.
4. F1. Feed rate in ipm for the plunge cut into the pocket.
5. F2. Feed rate in ipm for the general pocketing metal removal.
6. F3. Feed rate in ipm for the finish cut around the pocket.
7. 0. Offset override indicator. Either 0 or 1. If 0 is used, the computer will check for uncut areas that may be missed by using the offset specified. If the computer check is not desired, use 1.
8. P. Point type. If this number is 0, the pocket must be specified by cutter-center locations at the pocket corners. If the number is 1, the pocket is specified by its actual corners or, rather, the corners resulting from extending the pocket sides until they meet.
9. PT1, PT2, PT3, etc. These are the symbolic names or XYZ coordinates of the points that define the pocket, at the bottom

plane of the pocket. The cutting sequence follows the order in which the points are programmed, and the Z values define the pocket bottom.

The PØCKET routine will be illustrated here by programming the large pocket in the base pad, part print no. 2, though this is the simplest of all possible pocket shapes.

The lower left-hand corner of the work piece is given the coordinates X10 Y06, with +X to the right and +Y upward on the drawing.

PØCKET Data

1. The cutter is a two-lipped end mill or slot drill 1 in. in diameter. An effective radius of 0.4 in. is arbitrarily selected.
2. Cutter offset is taken as 1.5 in. (offset between passes divided by actual cutter radius).
3. Depth of finish cut at wall divided by cutter radius is 0.06 in. (0.03/0.5). The actual finish cut will be 0.030 in.
4, 5, 6. The several feed rates are 8, 16, 10 ipm.
7, 8. Offset override indicator and point type are 1. The pocket is specified by its actual corners.
9. The pocket corners are (11, 10), (11, 16.5), (19, 16.5), (19, 10).

Program

```
PARTNØ   PØCKET MILL/TØØLING PLATE
NØPØST
CLPRNT
CUTTER/1.0
REMARK   LØWER LH CØRNER X10 Y06
SETPT = PØINT/5,16,8
PTA   = PØINT/11.5,14.5,1
FRØM/SETPT
GØTØ/PTA
PØCKET/0.4,1.5,0.06,8,16,10,1,1,11,10,0,11,16.5,0, $
         19,16.5,0,19,10,0
GØDLTA/0,0,1
GØTØ/SETPT
FINI
```

PTA is selected to be inside the pocket, so that when the cutter cuts a ramp down to the pocket bottom no part of the pocket wall will be inadvertently removed. The cutter is withdrawn in Z before moving off the pocket area.

CLPRNT

The computer tabulates the tool positions in X, Y, and Z. The following are the XY positions after PTA in the printout:

	X	Y
1	15.845	13.345
2	14.155	13.345
3	13.780	12.780
4	13.780	13.720
5	16.220	13.720
6	16.220	12.780
7	13.780	12.780
8	13.030	12.030
9	13.030	14.470
10	16.970	14.470
11	16.970	12.030
12	13.030	12.030
13	12.280	11.280
14	12.280	15.220
15	17.720	15.220
16	17.720	11.280
17	12.280	11.280
18	11.530	10.530
19	11.530	15.970
20	18.470	15.970
21	18.470	10.530
22	11.530	10.530
23	11.500	10.500
24	11.500	16.000
25	18.500	16.000
26	18.500	10.500
27	11.500	10.500

This would be a sufficiently complex sequence to program manually, even though the pocket shape is quite simple. The computer charge for this printout in 1971 was $11.60, or 43¢ per tool position. A plot of the pocketing tool path is given in Fig. 152.

14.4. PATERN/ FOR HOLE PATTERNS

The PATERN routine is a positioning command like GØTØ and GØDLTA. It allows the programmer to set up patterns of hole locations

Patern/ for Hole Patterns 267

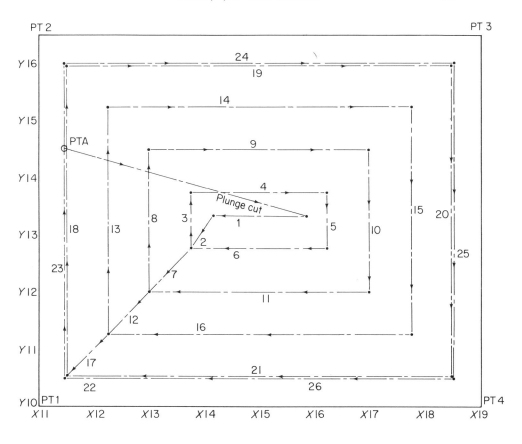

Figure 152

or points, to a maximum number of a few hundred depending on the APT system, in the following configurations:

    Evenly spaced line of points PATERN/LINEAR
    Circular pattern of points   PATERN/ARC
    Grid of holes or points   PATERN/PARLEL
    Random pattern of points   PATERN/RANDØM

  Pattern definitions will usually be programmed for machining operations in the following typical sequence:

    PATA = PATERN/LINEAR
    CYCLE/DRILL
    GØTØ/PATA
    CYCLE/ØFF

The postprocessor command CYCLE/DRILL will call up the required G81 or other drill function for each hole in the pattern. TAP, BØRE, or other operations may be required at each hole in the series.

1. PATERN/LINEAR

This type of pattern is defined by the first hole, the last hole in the hole sequence, and the total number of holes, thus:

PAT1 = PATERN/LINEAR,PT1,PT2,5

The holes must be evenly spaced. See Fig. 153. There are other formats

Figure 153

for this definition, but these will not be discussed.

2. PATERN/ARC

All the holes or points in this pattern definition lie on a circular arc. The pattern is defined by a previously defined circle, the angle to the X axis of the first point, the angle of the last point, the direction of rotation, and the total number of evenly spaced hole locations. See Fig. 154.

PAT2 = PATERN/ARC,CIR1,15,270,CCLW,7

Angles are given in degrees and decimal degrees as usual in APT. Figure 155 shows a pattern defined clockwise with a negative angle. A variant of the PATERN/ARC allows unevenly spaced holes to be defined, but is not discussed here.

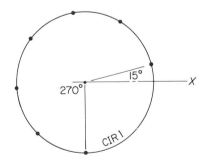

Figure 154

3. PATERN/PARLEL

Two linear patterns are defined, the two forming the basis of the grid or PARLEL pattern. See Fig. 156. A circular pattern is allowed instead of a linear pattern in the grid. The two subpatterns must include a common point as either their first or last point.

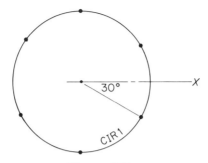

Figure 155

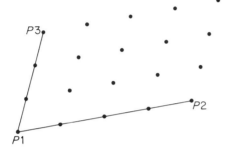

Figure 156

4. PATERN/RANDØM

This is a number of point locations not necessarily located on a straight or circular line. See Fig. 157.

Figure 157

PAT6 = PATERN/RANDØM,P1,P2,P3,P4

The random pattern may include a previously defined pattern:

     PAT7 = PATERN/RANDØM,P1,P2,P3,P4,PAT1
or   PAT7 = PATERN/RANDØM,PAT1,P1,P2,P3,P4
or   PAT7 = PATERN/RANDØM,PAT1,(PØINT/x,y),(PØINT/x,y),
           (PØINT/x,y), etc.

The PATERN/ routine processes remarkably little information. Cannot a smaller and cheaper computer be used for these simpler programming routines? The answer, of course, is yes. The APT language can be modified and some of its more complex computing routines surrendered, or a different language, better suited to simpler programs, may be used. But even the problem of processing complex APT programs on a small computer has been solved. These topics are the subject of the following chapter.

15

SMALL COMPUTERS AND SPECIAL LANGUAGES

15.1. ADAPT

Much numerical-control programming is limited to simple positioning or contouring in two dimensions and may not require the extensive computing and storage capacity of a large APT computer. The programming of the much-discussed O-ring groove is such a simple two-dimensional program, which makes only limited demand on computer capacity.

An obvious method of executing APT routines on smaller computers would be to restrict the range of programs to simpler two-dimensional types, but to retain the APT vocabulary and statement structure. This is the principle of ADAPT (ADaptation of APT), which can be run on a computer with 40K to 64K of word storage. ADAPT, therefore, is a subset of APT. The third-axis capability of ADAPT is restricted to such cases as sloped planes, ZSURF, and others, but it is incapable of handling the more complex shapes of the following chapter. The previous chapters have in general been an explanation of both APT and ADAPT programming, both of which may be run on the same computer system.

A more extensive discussion of ADAPT cannot be given here. Capability and programming details vary somewhat, depending on the computer used. The differences in programming required by different APT computer systems are by comparison quite minor.

15.2. UNIAPT

Only the very largest manufacturing companies can justify the purchase of a computer of sufficient size to handle APT programs, and few of these large companies could keep such a computer occupied with APT programs only. Smaller companies must send their programs to a computer service. This would not be a serious inconvenience or delay if the APT program were correct on the first try. But APT programs of any complexity are not usually satisfactory on the first try—the computer may be unable to find a point of tangency, for example (see Question 11 at the end of Chapter 12). The program, therefore, may have to make two or more trips to the computer terminal, and it is possible that a week may be lost between the first trial and the final acceptable tape.

Reducing APT to the more restricted ADAPT is not always an acceptable solution. The same programming difficulties may arise. The wide programming scope of APT is surrendered, and a 40K computer capacity for ADAPT is still beyond the scope of most companies.

Most companies using numerical-control methods could justify the purchase of a computer costing several thousand dollars. If the computer were on site, the programmer could try out his program, make corrections immediately, and return the program to the computer in a few minutes. Fortunately, a method of processing APT on small computers was invented a few years ago by United Computer Corporation of Carson, California. Their APT processor system, called UNIAPT, can be applied even to a computer as small as the desk-top PDP-8 of Digital Equipment Corporation. Compared to a large computer with APT capability costing about $10 a minute, the UNIAPT computer and software would cost about $10 an hour. The DPP-8/1 computer with peripheral equipment currently costs about $25,000 and the UNIAPT software about $12,000.

The language and statement structures of UNIAPT are the same as those used for APT. The UNIAPT system currently embraces almost the whole range of APT routines, though it has limited four-axis capability. Postprocessors are provided for a wide range of n/c machines, and the UNIAPT system may even be incorporated into a system for direct computer control of n/c machines.

The UNIAPT software package consists of four major components:

1. The United Programming Language Disk System, UPLDS
2. The UPL Interpreter
3. UNIAPT System Processor
4. UNIAPT Postprocessors

15.3. QUICKPOINT

Digital Equipment Corporation has devised a programming method and language for use on their small PDP-8 computer, called Quickpoint-8. This is a positioning language for such operations as drilling and tapping, but with simple programming tricks it is possible to output curves and lines at any angle. Only a few commands are used, and the language, therefore, is quickly learned. The commands are entered into the computer by typewriter.

Fig. 158. QUICKPOINT, a general-purpose computer system for n/c tape preparation, by Digital Equipment Corporation. The new system uses a conversational language and reduces greatly the time required for tape preparation.

In Quickpoint language, absolute coordinates are entered by addressing the word:

X3.250 Y5.5 Z1.5625

Incremental coordinates are addressed with a D before X, Y, or Z:

DX1.00 DY0.5 DZ1.25

274   Chap. 15 / *Small Computers and Special Languages*

Negative numbers and negative increments must, of course, include the minus sign. Incremental and absolute coordinates may be mixed in the same statement:

DX1.00 Y4.0 Z0.0

All G, M, F, T and other functions are included in the information as they occur.

Quickpoint uses the following statements:

1. OFS/x,y. Offsets the origin of coordinates.
2. MOV/x,y. Locates the starting point of any geometric command. Provides center coordinates for circles, arcs, bolt hole circles, etc. No machining operations are performed at the MOV location.
3. BHC/. Specifies bolt holes on a circle by radius of circle, counterclockwise angle from X axis of the first hole, and number of holes, in that order. Thus,

BHC/3.5 0 5

The bolt hole coordinates are computed counterclockwise. The starting angle may be negative. An example is given in Fig. 159,

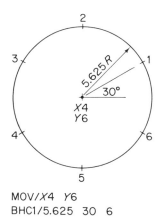

MOV/X4 Y6
BHC1/5.625 30 6

Figure 159

where the BHC is given the identifier 1 (BHC no. 1).

4. ARC/. Specifies holes along an arc of 360 deg or less by radius, angle from X axis of the first hole, incremental angle between

holes, and number of holes, in that order. The two angles may be positive or negative. See Fig. 160.

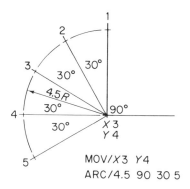

MOV/X3 Y4
ARC/4.5 90 30 5

Figure 160

5. LAA/. This statement (Line At Angle) allows incrementing along a line at an angle to the X axis by specifying the incremental distance, the angle to the X axis, and the number of holes. The angle may be positive (counterclockwise) or negative (clockwise). An example is given in Fig. 161. Note that

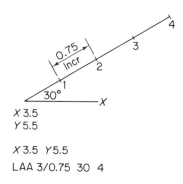

X 3.5 Y5.5
LAA 3/0.75 30 4

Figure 161

the starting point must be given in a preceding statement and that this point is not included in the number of points.

6. INC/ (Increment). This command replaces LAA when incrementing is along a line parallel to either the X or Y axis. The starting point is given in a preceding statement and is not included in the number of points. The command requires the specification of direction (R for right, L for left, U for up, D

for down), increment distance between holes, and number of holes.

Two INC patterns are shown in Fig. 162. The following is one of several possible programs for the vertical pattern of points.

X5 Y10	P1
DX−2.5 DY1.5	P2
DX1.5 DY2.0	P3
DX1.5 DY1.0	P4
INC1/D 1.0 4	vertical pattern

The following is one of several possible programs for the horizontal pattern of holes.

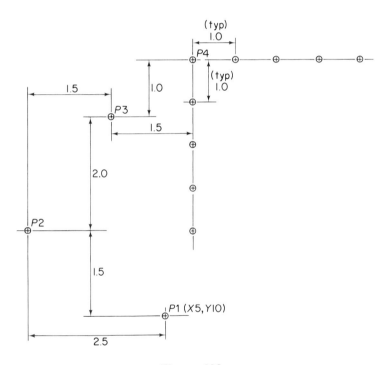

Figure 162

OFS/X5 Y10	
X0 Y0	P1
DX0.5 DY4.5	P4
INC2/R 1.0 4	horizontal pattern

If P4 were defined by a MOV command, then no machining operations would be performed at P4.

7. GRD/ (Grid). This statement computes a pattern of holes parallel to X and Y axes. A starting point must be specified in the previous command, and this point may be located at any corner of the grid. The direction of the pattern, the increment, and the number of holes are specified for both of the patterns of the grid, in that order. The starting point is not counted in the number of holes. In Fig. 163 a 5 × 4 hole pattern is specified.

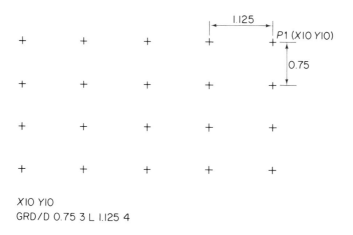

Figure 163

15.4. CONTOURING WITH QUICKPOINT

Though Quickpoint is a positioning language, it is possible to use it for simpler types of contouring in linear interpolation. The contouring methods discussed here may, of course, be employed with any other positioning language, such as AUTOSPOT.

To program a line at an angle for a positioning mill restricted to X and Y increments, the LAA command for a series of holes is used. The required incremental distance and number of steps to mill the line are determined (see Sec. 4.2). Suppose that to mill the line-at-angle, 40 increments in X and Y are required on a line at 60 deg counterclockwise to the X axis. The command would be

LAA/.006 60 40

The 40 points thus computed are used as milling positions for the cutter.

The same procedure is used for contouring a circular arc. Suppose that the same positioning mill must mill an arc 1.0 in. in radius with a

0.25-in. end mill over 90 deg, as in Fig. 164. With tool offset, the radius is 1.125 in. Use the ARC/ command

MØV/X6 Y8
ARC/1.125 0 0.5 180

where 1.125 = radius of tool path
0 = angle of the starting point with respect to X axis
0.5 = desired angle between milling positions
180 = number of milling positions

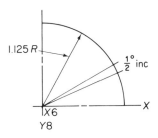

Figure 164

15.5. A QUICKPOINT PROGRAMMING EXAMPLE

Part print no. 7 shows the drilling pattern for the front tube sheet of a Scotch Marine boiler. For those unacquainted with the energy industry, perhaps it might be explained here that there is nothing Scotch about a Scotch Marine boiler, nor is it used in the power plant of ships. This is the type of boiler that heats schools and commercial buildings. This tube sheet will be used to illustrate the Quickpoint method. The program given below is not the best or the most efficient of all possible programs.

The two holes P1 and P2 toward the top of the tube sheet are programmed first. After drilling P2 a tool change is necessary before drilling the tube holes; therefore, an M06 tool change function is included. The lowest tube hole on the right-hand side, P3, is drilled first, then the row of two holes, the row of three holes, and the program example terminates with the lowest five-hole horizontal row.

X14 Y43.75	P1
DX—28 M06	P2
X21.3125 Y5.375	P3
DX1.9375 DY3.375	first hole of two-hole row
DX—3.875	second hole of two-hole row
DX—1.9375 DY3.375	start point, three-hole row
INC1/R 3.875 2	three-hole row

DX5.8125 DY3.375   start point, five-hole row
INC2/R 3.875 4   five-hole row

Since either word address or tab sequential format is required for tape programming, and for other reasons, postprocessors are required for Quickpoint.

15.6. AUTOSPOT

AUTOSPOT (AUTOmatic System for POsitioning Tools) is an IBM positioning language with the capability of simple two-dimensional contouring of circular arcs and lines at an angle, like Quickpoint. Like APT, the AUTOSPOT system is composed of a language and a processor, requiring a postprocessor to adapt the processor output to a specific numerically controlled machine. Postprocessors are called up by a MACHIN/ statement, as in APT-ADAPT. The particular computer used influences the AUTOSPOT statement format to some degree, and the following discussion refers to AUTOSPOT for the IBM 360 series of computers.

The statement formats for AUTOSPOT resemble in many ways those for APT. Statements may be inserted anywhere in card columns 1–72, with columns 73–80 for card numbers. PARTNØ, REMARK, and FINI have the same uses as in APT programming. In AUTOSPOT as in APT, a slash separates the major word from its detailed information. Commas are used to separate words when no other punctuation, such as parentheses, is available to separate words. In APT statements, only commas separate words. The single dollar sign indicates that the AUTOSPOT statement is continued on the next line. The double dollar sign precedes any commentary.

15.7. AUTOSPOT GEOMETRIC STATEMENTS

The geometric statements and the format allowable in AUTOSPOT are identical to those used in APT, though the range of statements is restricted. Symbolic names are selected by the programmer, subject to the same restrictions as obeyed by APT statements: not more than six characters, the first character cannot be a number, etc. Only XY coordinates are specified. Nested definitions are allowed. The allowable geometric statements are these:

    1. PØINT. Six allowable definitions.
       a. XY coordinates. PT1 = PØINT/3,4.
       b. Reference point, radius, angle:

          P1 = PØINT/3,10,RADIUS,5.1,ATANGL,60 (Fig. 165)

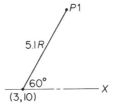

Figure 165

c. Intersection of two lines.

$$P1 = PØINT/INTØF, L1, L2$$

d. Intersection of a line and a circle.

$$P1 = PØINT/XLARGE, INTØF, L1, C1$$

e. Intersection of two circles.

$$P1 = PØINT/YSMALL, INTØF, C1, C2$$

f. Center of a circle.

$$P1 = PØINT/CENTER, C1$$

A more extended explanation of these definitions is given in Chapter 11.

2. LINE. Eight allowable definitions.
   a. Coordinates of two points.

   $$L1 = LINE/3, 4, 3.1, 8.9$$

   b. A point and angle to the X axis.

   $$L1 = LINE/3, 4, ATANGL, -30$$

   c. Passing through a point and tangent to a circle (Fig. 166).

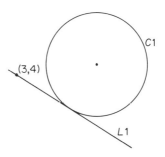

Figure 166

L1 = LINE/3,4,RIGHT,TANTØ,C1

d. Tangent to two circles (Fig. 167).

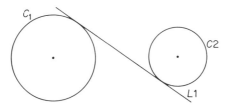

Figure 167

L1 = LINE/LEFT,TANTØ,C1,RIGHT,TANTØ,C2

e. Through a point at an angle to another line.

L1 = LINE/3,4,ATANGL,45,L2

f. Through a point and parallel to another line.

L1 = LINE/3,4,PARLEL,L2

g. Through a point and perpendicular to another line.

L1 = LINE/3,4,PERPTØ,L2

h. Parallel to another line at a distance normal to both lines.

L1 = LINE/PARLEL,L2,YSMALL,1.5

3. CIRCLE. Five allowable definitions.
   a. Center coordinates and radius.

   C1 = CIRCLE/CENTER,3.2,4.2,RADIUS,3.5

   b. Tangent to two lines with specified radius (Fig. 168).

   C1 = CIRCLE/YLARGE,L1,YSMALL,L2,RADIUS,2.0

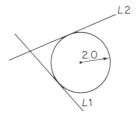

Figure 168

c. Tangent to a line and a circle with a specified radius (Fig. 169).

C1 = CIRCLE/YSMALL,L1,XSMALL,ØUT,C2,RADIUS,2.0

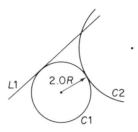

Figure 169

d. Tangent to two circles with a specified radius (Fig. 170).

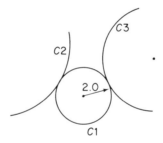

Figure 170

C1 = CIRCLE/TANTØ,ØUT,C2,ØUT,C3,YSMALL,RADIUS,2.0

e. Tangent to a line with specified center coordinates.

C1 = CIRCLE/CENTER,4,4,TANTØ,L1

15.8. TYPES OF AUTOSPOT STATEMENTS

Four types of statements are employed in AUTOSPOT programming:

1. Geometric statements, just discussed.
2. Specification statements.
3. Pattern statements.
4. Machining statements.

Each type of statement has its own format.

Geometric Specification	Symbolic name = major word/(definition details)
	Word = (value)
Pattern	Symbolic name = PATERN/(locations)
Machining	Operation/(modifiers)/(locations)

15.9. SPECIFICATION STATEMENTS

Specification statements are modal; that is, they are in force through the program until modified by another similar specification. There are eight specification statements.

1. TP. Table position. Applies to machines with rotary tables, such as the Milwaukee-Matics. TP(0) indicates the zero position of the table, TP(4) position 4.

2. SAFE. Defines a location to which the cutter is withdrawn before a rotary table is rotated. Thus SAFE = 2.0,8.0,3.0 calls for a SAFE position at those coordinates.

3. CL. Clearance amount above the work. The tool will be withdrawn this amount above the part when in rapid traverse from point to point. This distance must, of course, be high enough to avoid clamps or other obstructions, unless the tool is to be programmed around them. Thus CL = 0.375 calls for $\frac{3}{8}$-in. clearance.

4. DH. Deep-hole drilling sequence. The woodpecker drilling sequence is defined in terms of drill diameter:

    $$DH(1) = 3.0, 1.5, 1.0$$

    This statement is the first (1) DH statement. The tool is to drill three diameters, then retract to clear chips. The tool will then drill out another $1\frac{1}{2}$ diameters of hole, retract, then drill one diameter. If the full depth of hole has not been reached, the last depth specification (1.0) will be repeated until the required hole depth is produced. The required depth of hole is not given in the DH statement.

5. XYØNLY. This statement indicates that the AUTOSPOT processor is to suspend interpretation of tool and machining statements and Z depths. The part program is taken as two-dimensional, though all three-dimensional information is retained for postprocessing.

6. BKC. Break chips. The rotation of the cutter is stopped for chip-breaking. Similar to DH.

    $$BKC = 3.0, 1.5$$

The statement specifies that the cutter is to be stopped at 3 and 1.5 diameters. The cutter does not retract.

7. DA. To be discussed below.
8. TØØL. To be discussed.
9. TØLER. Specifies a tolerance value if required by the postprocessor. Thus TØLER/.005.

15.10. DATUM POINTS

The zero point or origin of the n/c machine must be that of the part program. However, the problem of converting part coordinates to the machine coordinate system is eased by the use of the DA (dash) statement. This is explained by reference to Fig. 171.

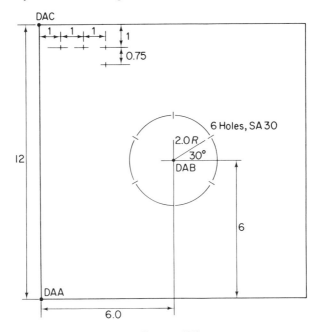

Figure 171

The lower left-hand corner of the part is referenced DAA. The statement DAA = 3.0,−4.5,6.2 states that DAA is a point on the part that is +3.0 in. in X from the machine origin, −4.5 in. in Y, and +6.2 in. in Z. The third letter (A) in DAA is the programmer's choice; it may be any letter (X, Y, Z are not allowed when programming for the IBM 1620). The Z offset in the Dash statement may be omitted for two-axis machines. Part coordinates may be stated with reference to DAA as a substitute origin, leaving the AUTOSPOT processor to calculate true coordinates.

It is easier to program a series of holes in a bolt circle from the center of the circle. The bolt circle in the figure may be programmed from its center by means of two Dash statements:

$$DAA = 3.0, -4.5, 6.2$$
$$DAB = DAA(6.0, 6.0, 0.0)$$

The second statement, DAB, references DAB with respect to DAA, as DAA does to the machine origin. Thus with a Dash statement the working origin may be moved to any point of convenience. For the three holes at the top of the work piece, DAC is used for determining their coordinates. DAC may be referenced to DAA or DAB.

$$DAC = DAA(0.0, 12.0, 0.0)$$
$$DAA = 3.0, -4.5, 6.2$$

or

$$DAC = DAB(-6.0, 6.0, 0.0)$$
$$DAB = DAA(6.0, 6.0, 0.0)$$
$$DAA = 3.0, -4.5, 6.2$$

For a machine with a rotary table, the table position must also be specified in the Dash statement:

$$DAF = 12, 12, 4.6, TP(2)$$

15.11. PATTERN DEFINITIONS

Pattern definitions may be included within machining statements. Here we discuss a primary form of pattern definition, with the format

(name) = PATERN/(point specification)

1. **Absolute Dimensions.** Consider the four top holes of Fig. 171. This hole pattern may be thus described:

PAT1 = PATERN/DAC(1.0,−1.0)(2.0,−1.0)(3.0,−1.0)(3.0,−1.75)

with DAC previously defined.

If this pattern should be incorporated into a machining statement, the format is

PAT1 = DRILL,110/DP(1.0)/DAC(1.0,−1.0)(2.0,−1.0)(3.0,−1.0)(3.0,−1.75)

The statement specifies that the pattern is to be drilled with tool no. 110 to a depth of 1.0 in.

If points are to be machined at different heights on the work piece, then the point coordinates within the parentheses should include Z coordinates also.

2. **Incremental Dimensions.** Points in a pattern may be specified by

incremental dimensions from a previous point. If the four top holes of Fig. 171 are used as an example, the incremental pattern statement would be

PAT1 = PATERN/DAC(1.0,−1.0)DELTA(1.0)DELTA(1.0)DELTA(0.0, −0.75)

The first increment in the parentheses is delta X, the second delta Y, and the third, if present, delta Z. If only X is incremented, only this number need appear.

3. **End Points.** Linear patterns may be specified by end points and number of points. Figure 172 is used as an example. For this

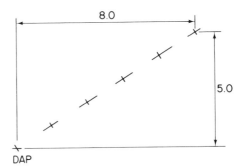

Figure 172

pattern the following statement applies:

PAT2 = PATERN/DAP,SX(0)SY(0)EX(8.0)EY(5.0)NH(6)
SX = starting X coordinate with respect to DAP
SY = starting Y coordinate with respect to DAP
EX = ending X coordinate
EY = ending Y coordinate
NH = number of holes or points.

4. **Other linear pattern statements** are available, but are not discussed here.

5. **Bolt Hole Circles and Circular Arrays.** The format for these is the following, if the bolt hole circle of Fig. 171 is used as an example:

PAT3 = PATERN/DAB,AT(0,0)RADIUS,2.0,SA(30)IA(60)NH(6)

The pattern is located with reference to DAB. The center of the circle is given by AT (i.e., at DAB). The radius is 2.0 in., the starting angle SA is 30 deg., the incrementing angle is 60 deg., and the number of holes is six.

The reference point AT may be specified also by XYZ coordinates, or by a symbolic name, such as PT1.

15.12. MACHINING STATEMENTS

Machining statements are available in AUTOSPOT for both positioning and simple contour milling operations.

In positioning operations, the tool positions to clearance (CL) above each point programmed, cuts to the depth specified, retracts to CL, then moves to the next point programmed. In milling operations, the tool positions to CL above the starting point of the milling operation, plunges to the specified depth at feed rate, then mills the path programmed. This milling path may include contouring of lines and arcs.

The structure of a machining statement was given in Sec. 15.11 in the discussion of absolute dimensions. The operation section of the statement gives the machining operation to be executed, the tool number, and sometimes special modifiers such as TLLFT, TLRGT. TLLFT and TLRGT have the same meaning as in APT statements, with offset equal to tool radius. The more common machining operations are the following (other machining words are also used in AUTOSPOT programming):

DRILL	drill	PMILL	pocket mill
SPDRL	spot drill	FMILL	face mill
CSK	countersink	FLF	face mill to the right of the left boundary
BORE	bore		
CBORE	counterbore	FRT	face mill to the left of the right boundary
REAM	ream		
TAP	tap	FUP	face mill below the upper boundary
PUNCH	punch	FDWN	face mill above the lower boundary
MILL	mill	PILOTD	pilot drill

The second part of the machining statement contains modifiers that control the machining process. These modifiers may be in any order that the programmer chooses and in any number of modifiers. The more common of these modifiers are

DP	depth	DH	deep hole
DI	diameter	SS	spindle speed
FR	feed rate	DW	dwell
CL	clearance	ONKUL	coolant on
BKC	break chips	OFKUL	coolant off

Details for each of these words follow the word in parentheses.

Some examples of positioning statements will make the AUTOSPOT structure and meaning clear.

1. DRILL,0150/DP(1.0)/PAT1
   Drill no. 0150 is to drill the previously defined PAT1 to a depth of 1.0 in.
2. REAM,1100/DP(2.6)/DAA(3.0,4.0)(5.5,6.5)(8.0,8.0)
   Reamer no. 1100 is to ream the three holes given, all to a depth of 2.6 in. Coordinates are with respect to point DAA.
3. DRILL,0065/DP(1.65)/DAB(3.0,3.0)(5.0,3.0,−1.5)
   See Fig. 173. The second position includes a Z coordinate because this drilling location is 1.5 in. below the plane in which lie DAB and the first point.

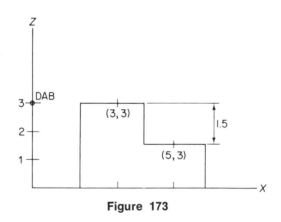

Figure 173

4. DRILL,0065/DP(1.65)/DAB(3.0,3.0)(2.0)(5.0,3.0)
   See Fig. 174. The section C of the part will cause a tool collision

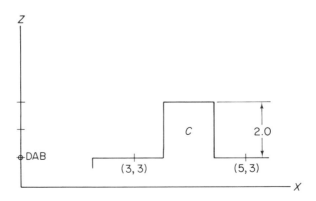

Figure 174

Machining Statements

when the tool moves from the first point to the second. The word (2.0) retracts the spindle by this amount to clear the obstruction.

5. SPDRL,0650/DP(0.25)/DAB,AT(2.0,3.0)RADIUS(2.0)
   SA(30)IA(60)NH(3)

   Spot drill no. 0650 is to drill 0.25 in. deep. The hole pattern lies on a circular arc of radius 2.0 in. with center at (2, 3) from DAB. The starting angle is 30 deg counterclockwise from the X axis, and each hole is incremented another 60 deg for a total of three holes.

6. DAK = 3.0,6.0
   DH = 3.0,1.5
   DRILL,16/DP(3.0)DH/DAK,SX(2.0)SY(4.0)EX(5.0)
   EY(1.0)NH(4)

   The drill must drill 3 in. deep. Let us assume a $\frac{1}{2}$-in. drill for drill no. 16. The DH sequence calls for the drill to enter 3 diameters or $1\frac{1}{2}$ in., then retract, then to drill $1\frac{1}{2}$ diameters or $\frac{3}{4}$ in. further. It is to repeat the advance of $1\frac{1}{2}$ diameters until the required depth is reached (the last advance need not be exactly $1\frac{1}{2}$ diameters). The starting X and Y positions of the pattern are (2, 4), the ending X and Y positions are (0, 7.625), and there are four points in the pattern including the two end points.

   An alternative drilling statement is the following:

   DRILL,16/DP(3.0)DH/DAK,SX(2.0)SY(4.0)DX(1.0)DY(1.0)NH(4)

7. A pattern may be defined in a machining statement and then reused in other machining statements. The three holes (1.0) (3.0),(2.0,3.0), and (4.0,4.0) with respect to DAD are to be spot drilled, drilled, and reamed. The pattern is defined only for the first operation.

   PAT2 = SPDRL,67/DP(0.20)/DAD(1.0,3.0)(2.0,3.0)(4.0,4.0)
   DRILL,94/DP(0.6)/PAT2
   REAM,106/DP(1.1)/PAT2

   The pattern is defined in the location section of the first statement, and only that portion of the statement is associated with the symbolic name PAT2. Alternatively, the hole pattern could be defined in a separate pattern statement:

   PAT2 = PATERN/DAD(1.0,3.0)(2.0,3.0)(4.0,4.0)
   SPDRL,94/DP(0.6)PAT2
   DRILL — — etc.

8. Several patterns may be called up in a machining statement:

DRILL,94/DP(0.75)/DAD,PAT2,DAE,PAT3

where PAT2 is referenced to point DAD and PAT3 to DAE. If both are referenced to DAD, then

DRILL,94/DP(0.75)/DAD,PAT2,PAT3

9. Points and patterns may be mixed:

DRILL,94/DP(0.75)PAT2,DAF(4.0,4.7)(5.6,5.9)

If a symbolic pattern name is followed by point specifications, as in this statement, then a DA reference must intervene between them.

15.13. PATTERN MANIPULATIONS

1. A pattern may be reused at another location by specifying the new starting point of the pattern. Suppose PAT1 of Fig. 175 is

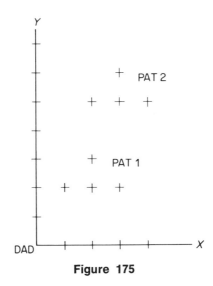

Figure 175

to be repeated in PAT2. PAT1 is first defined:

DAD = 3,8,0
PAT1 = PATERN/DAD(1,2)(2,2)(3,2)(2,3)

To translate PAT1 to the second location, use the statement

PAT2 = PATERN/DAD,PAT1(2,5)

The ordering of the points is retained in the translated pattern. There are two requirements for such translation: the new starting point for the pattern (2, 5) must be included, and the DA reference point must precede the new pattern even if this reference is unchanged, as it is in this example.
2. Patterns may be machined in reverse order of points by the use of the word REV. Thus,

DRILL,160/DP(0.5)/REV,PATA

calls for drilling the pattern from the last defined point to the first defined point.
3. Patterns may be repeated by incrementing. Consider the grid of holes of Fig. 176. The module from which this grid is built

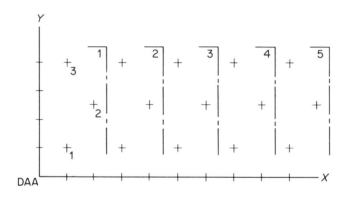

Figure 176

contains the three holes numbered 1, 2, 3. Define this module as PAT1.

      PAT1 = PATERN/DAA(1,1)(2,2.5)(1,4)
and   PAT2 = PATERN/DAA,PAT1,SX(1)SY(1)EX(9)EY(1)NR(5)

The DAA may be omitted, since it is included in the statement for PAT1. SX and SY give the coordinates of the first point of the first pattern repetition and EX and EY of the first point of the last pattern repetition. NR means number of repetitions. DX and DY may be used instead of EX and EY.

The repeated pattern may be defined in a machining statement thus:

DRILL,66/DP(0.6)/DAA,PAT1,SX(1)SY(1)DX(2)NR(5)

Similarly, the grid of holes in Fig. 177 may be defined as

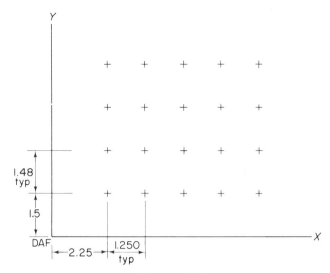

Figure 177

follows by means of a drill statement:

PAT1 = PATERN/DAF,SX(2.25)SY(1.5)DX(1.25)NH(5)
PAT2 = DRILL,102/DP(0.5)/DAF,PAT1,SX(2.25)SY(1.5)DY(1.48)NR(4)

4. Other pattern manipulations are available in the AUTOSPOT system.

Questions

1. Set up pattern statements for all the tube holes on the left-hand side of the tube sheet of part print no. 7.
2. These tube holes are to be spot drilled, drilled, and reamed. Set up a complete program for these machining operations.

15.14. POSTPROCESSING OF AUTOSPOT PROGRAMS

In general, the postprocessor vocabulary for AUTOSPOT is that of APT-ADAPT. The postprocessor is called up by the proper MACHIN/ statement. If there is to be no postprocessing, then NØPØST applies as in APT. PARTNØ, FINI, and CLPRNT also are used in AUTOSPOT.

A partial postprocessor vocabulary, applicable also to AUTOSPOT, was supplied in Chapter 13.

15.15. TOOL STATEMENTS

Tool statements are not required in any program that does not specify a tool number in the machining statement, if the specification statement XYONLY is used in the program.

The information in the tool statement is used by the computer to determine such machining details as tool offsets and drilling depths. The significant dimensions of the tool are shown in Fig. 178, where

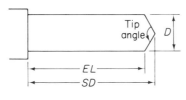

Figure 178

D = diameter
EL = effective length
SD = setting distance

The format of the tool statement on the punched card is tabulated as follows.

Card Columns	Information
1–5	TOOL/
6–11	Operation, e.g., DRILL, SPDRL, etc. This entry is optional.
12–16	Tool number
17–23	Tool diameter. If no diameter is specified, the computer assumes a zero diameter.
24–29	Tip angle of tool in degrees. If not specified, it is taken to be zero.
30–36	Setting distance or total length of tool and holder which extends beyond the Z = 0 position. If not specified, this dimension is assumed to be zero.
37–43	Effective length. Normally the chamfered end of the tool is not considered when one is computing drilling depths. If effective length is not specified, the setting distance is used.
44–47	Spindle speed in rpm. This is the normal operating speed of the tool and can be changed by a later postprocessing statement or an entry as a modifier in a machining statement.

48	Direction of rotation. L indicates counterclockwise; no entry indicates clockwise.
49–54	Normal feed rate. If not specified, feed rate is assumed to be zero. The feed rate may be varied by an entry as a modifier in a machining statement or by a postprocesser command.
55–56	Coolant code. ØN, ØFF, TK (tapkul), FL (flood), or MI (mist). If no coolant code is specified, ØFF is assumed. Actual M function numbers may be used.
57	Tool change operation. A or blank—automatic tool change K—key M—manual tool change
58–72	These card columns are blank. Comments may be inserted.
73–80	In accord with the usual practice, these columns are used for card identification numbers.

Questions

The work piece shown in Fig. 15-1 is taken from the IBM publica-

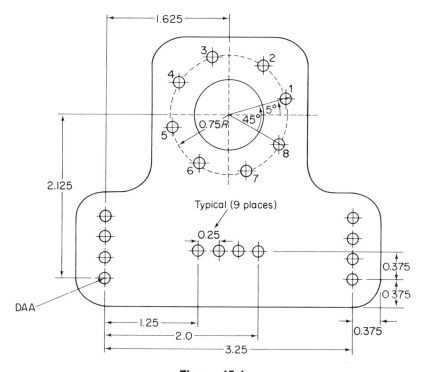

Figure 15-1

tion H20-0373-2, System/360 AUTOSPOT. Write the AUTOSPOT program for spot drilling and drilling the hole patterns of the part, given the following information:

1. *Tool information.* Spot drill #0646, diameter 0.100 in., 118-deg angle, setting distance 4.020, effective length 4.014, spindle speed 1200 rpm clockwise, feed rate 2.0 ipm. All tool changes are manual.
    Drill #0749, diameter $\frac{9}{32}$ in., 118-deg angle, setting distance 4.600, effective length 4.517, spindle speed 520 rpm clockwise, feed rate 2.5 ipm.
2. Clearance height = 0.200 in.
3. DAA = 4.0, 6.0
4. *Patterns.* In defining the three four-hole patterns, define first the horizontal pattern of four holes. Define the left-hand vertical pattern thus: PAT(0,0)ATANGL(90), where PAT is the previously defined horizontal pattern. Define the right-hand vertical pattern thus: PAT(3.25,0)ATANGL(90), where PAT is previously defined.
5. The program sequence should be

    PARTNØ
    CLPRNT
    TØØL/
    TØØL/
    CL =
    DAA =
    —
    FINI

15.16. MILLING OPERATIONS

1. MILLING WITHOUT TOOL OFFSET. There is no tool offset when one is milling grooves such as the O-ring groove. Such on-path milling is called up by a MILL statement with the general format

   MILL,(tool number)/DP(n)/DAA(starting point)path of tool.

Suppose that the tool must mill the path of Fig. 179 starting at point A, going then to B, C, D, and A. This series of milling cuts is executed by means of the statement

   MILL,M66/DP(0.375)/DAA(3,5)UTO(8.0)RTO(8.0)DTO(5.0)LTO(3)

Here UTO means "Mill up to."

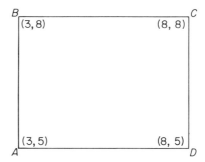

Figure 179

DTO means "Mill down to."
RTO means "Mill right to."
LTO means "Mill left to."

Note that only one coordinate is needed after each of these direction commands, since all motions are parallel to either X or Y and one coordinate does not change.

Sloped lines and circular arcs may be programmed in AUTOSPOT language. An example is shown in Fig. 180. The tool path is specified thus:

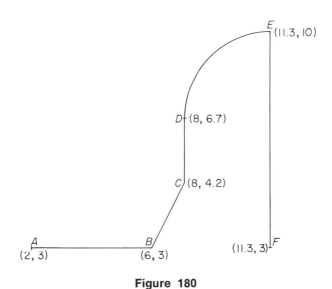

Figure 180

MILL,M66/DP(0.375)/DAA(2,3)RTØ(6,3)STØ(8,4.2)UTØ(6.7) $
CWARC(90)TØ(11.3,10)RADIUS(3.3)DTØ(11.3,3)

The starting point must, as usual, be programmed after the datum reference. This starting point need not be a corner. STO means "slope to" and CWARC means "clockwise arc." CCWARC means "counterclockwise arc."

In these contour milling operations the starting point must always be programmed. If the surface to be machined is at a different Z level than the datum reference, the starting point should include a Z value also. FR (feed rate) and SS (spindle speed) may be included in the machining details with the required data in parentheses.

If a depth is not stated, zero depth is assumed. If a depth is specified, the tool positions at clearance level above the starting point. Then it feeds to depth at one-half the specified feed rate and machines the path at feed rate. The command MILL, however, does not command a retraction to clearance at the end of the path. A retraction must be executed by any of a number of commands, including GØTØ (as in APT-ADAPT), STØP, or a ZTØ command, this latter being executed at half feed rate. Still another method is to use a Z avoidance number at the end of the MILL statement thus:

... /DP(0.375) ... CWARC(90)TØ(11.3,10)RADIUS(3.3)DTØ(11.3,3)(.75)

The depth is 0.375, and the Z avoidance 0.75. Therefore, the tool retracts to clear the surface by 0.75 in.

The milling path may also be defined by a previous pattern, by symbolic names for points, or by XY or XYZ values. If no depth is specified, the AUTOSPOT processor assumes the starting point to be off the part and positions to the starting point at a rapid rate with no allowance for clearance.

2. FILLETS. To mill a filleting arc between two intersecting lines, insert FIL(radius) between the two lines. Fillets between arcs and lines or arcs and arcs are also acceptable. The milling operation of Fig. 181 is programmed as follows:

MILL/DAA(2,2)RTØ(4)FIL(0.375)STØ(5,3.2)

Note that the terminal points (intersection points) of lines are specified as they would occur if there were no fillet.

3. TOOL OFFSET. Suppose that the rectangular shape of Fig. 179 is the shape of the work piece to be cut. Then the modifiers TLLFT or TLRGT are used. Looking in the direction of tool motion, the tool must be offset on the left side of the finished contour in cutting from A to B. Then TLLFT is required in the MILL statement:

MILL,M66,TLLFT/DP(0.375)/DAA, etc.

The AUTOSPOT processor calculates the amount of tool offset and the required tool path from the tool card and the description of the tool path in the MILL statement.

If a rough-milling cut is to leave an amount for a finishing pass, this amount is programmed in parentheses after TLLFT or TLRGT:

MILL,M66,TLLFT(0.05)/DP(0.375)/DAA, etc.

This is the equivalent of the THICK/ command in APT. The finish cut statement might be

MILL,M67,TLLFT/DP(0.375)/DAA, etc.

At any point along such milling paths as have been discussed, a Z movement to another depth may be specified by inserting ZTØ (z) at the appropriate place. The quantity in parentheses is the new Z value relative to the Z of the datum reference. It may be plus or minus.

4. FACE MILLING. The command FMILL calls up a face milling operation in which a rectangular surface is milled all over to required depth. The cutter movements exceed the boundaries of the part surface during FMILL. Figure 182 is used as an example of the FMILL routine.

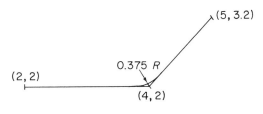

Figure 181

The statement format is this:

FMILL(tool)/DP(n)/DI(m)/DAA(starting point)tool path

The DI (diameter) word specifies the amount of stepover per pass as a fraction of cutter diameter. Thus for a 1.5-in. end mill, DI(0.5) calls for each pass across the work to be 0.75 in. from the previous pass. If DI is omitted, the stepover is the full tool diameter.

For the operation of Fig. 182, the required statement is

FMILL/DP(0.25)/DI(0.6)/DAA(2,1)RTØ(6.0)UTØ(4.)LTØ(2.0)

It will be noted that the shape of the part is specified, not the path of the cutter. The tool is initially positioned 0.6 × diameter from the part at the specified depth. It then mills across until its center is again 0.6 diameter from the other side of the part. It then steps over at rapid traverse, con-

Milling Operations

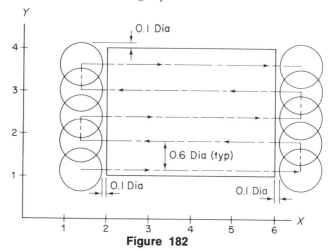

Figure 182

tinuing until completion of the operation. The width of the last cut is adjusted so that the tool overhangs the part by 0.1 diameter. All tool motions are parallel to X and Y axes of the machine.

5. POCKET MILLING. PMILL calls up pocket milling to a specified depth. The statement format is

PMILL(tool)/DP(n)/DI(m)/DAA(starting point)tool path

This is the same format as for FMILL.

The direction of milling is from the first point to the second point specified in the tool path. As in APT-ADAPT pocket milling, the cutter path is a quasi-spiral beginning near the center of the pocket and offsetting toward the periphery. The points of the path must be ordered in the desired sequence of cut. The first point must not be repeated as the last point; that is, a closed path must not be defined. As with the APT pocket routine, interior angles equal to or greater than 180 deg are not allowed. The path is described with the usual words UTØ, DTØ, LTØ, RTØ, and the points are the actual points of the pocket, not points on an offset tool path.

In entering the pocket, the tool positions at rapid traverse to clearance above the center of the pocket, then feeds at half feed rate to depth. The pocket is milled until the tool is one radius from the pocket sides. A finish amount may be allowed by means of the modifiers RIGHT(n) or LEFT(n). If DI is not specified, then DI is taken as 0.9 times tool diameter.

6. GØTØ. To avoid a clamp or other obstacle, a GØTØ command may be specified, with the format

GØTØ/DAA(point)

The point may have XY or XYZ coordinates. The point may have a symbolic name.

Questions

1. Program a FMILL operation over the whole face of the base pad, part print no. 2, using a 3.000-in. face mill with DI = 0.7. The lower left-hand corner of the base paid is to be affixed at X = 6, Y = 8.
2. Program the two pockets of the base pad. The lower left-hand corner of the part is affixed at X = 6, Y = 8. Rough-mill only.
3. Program the contour of part print no. 6, calibration block, in AUTOSPOT. Do not attempt to mill the small slot on the top surface. Use a 0.500 cutter, and affix the lower left-hand corner of the part at X = 3, Y = 10.

16

THE PROGRAMMER AS SCULPTOR

"Engineering design is becoming increasingly creative because of the ever-increasing sophistication of new products. Engineers will thus be more and more concerned with eliminating the routine and repetitive tasks. To allow time for greater creativity, there must be devices available to relieve the gifted person of much of the tedious routine which often overburdens him. The electronic digital and analog computer is the key to a new and more creative approach to engineering design." This quotation is taken from *Engineering Graphics and Numerical Control,* by R. B. Thornhill, published by McGraw-Hill Book Company.

Numerical control programming is only another useful mathematical technique. If APT is to be anything better than a transient technology, to be discarded in time as so many other engineering methods have been, it must expand, not restrict, the scope and creativity of the programmer. The intent of this and the following chapter is to indulge creativity by applying numerical control to more imaginative uses than it usually receives.

Above all else, creativity requires freedom. This chapter will present certain APT techniques that allow the designer greater freedom of shape in two and three dimensions. Hence the title of the chapter. It is not the intent to present the whole range of APT multidimensional routines, however. Only those of greatest scope in creative design are discussed; these would appear to be the QADRIC/, GCONIC/, and TABCYL/ routines.

16.1. QADRIC

This useful definition may be used to represent a range of two- and three-dimensional mathematical shapes. The general form of the QADRIC statement is

$$\text{QUAD} = \text{QADRIC}/a, b, c, f, g, h, p, q, r, d$$

where a, b, c, f, ... are the coefficients of the second degree equation

$$aX^2 + bY^2 + cZ^2 + fYZ + gXZ + hXY + pX + qY + rZ + d = 0$$

Consider the following examples.

1. A parabolic cylinder, $Y^2 = 3X$. See Fig. 183. The cylinder ex-

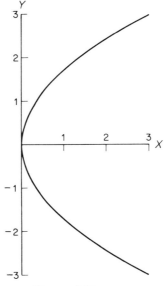

Figure 183

tends indefinitely in the Z direction. For this case,

$$a = c = f = g = h = q = r = d = 0$$

and PARCYL = QADRIC/0,1,0,0,0,0,3,0,0,0

2. The hyperbolic paraboloid, $X^2/a^2 - Y^2/b^2 = 2cZ$. This has the saddle shape of Fig. 184. For this case, the coefficients in the APT definition are

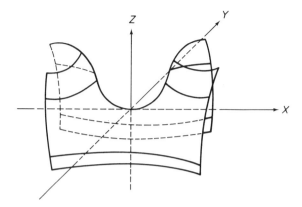

Figure 184

$$c = f = g = h = p = q = d = 0$$
$$a = b^2$$
$$b = a^2$$
$$r = 2a^2b^2c$$

and HYPAR = QADRIC/b²,a²,0,0,0,0,0,0,2a²b²c,0

3. Other quadratic surfaces with their standard equations are shown in Fig. 185.

16.2. GCØNIC

The QADRIC definition may be simplified for plane curves by reducing it to the GCØNIC:

GENCØN = GCØNIC/a,b,c,d,e,f

where a to f are the coefficients of the second degree equation

$$aX^2 + bXY + cY^2 + dX + eY + f = 0$$

This statement is handled in the same way as QADRIC. All zero coefficients must be included.

As an example of how to convert a normal mathematical expression to a QADRIC or GCØNIC statement, consider the parabola

$$(Y - a)^2 - k(X - b) = 0.$$

Expand: $Y^2 - 2aY + a^2 - kX + kb = 0$
Rearrange: $Y^2 - 2aY - kx + (kb + a^2)$
Then QPARA = QADRIC/0,1,0,0,0,0,−k,−2a,0,(kb+a²)

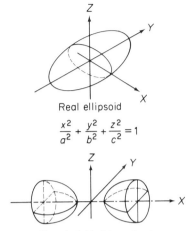

Real ellipsoid

$$\frac{x^2}{a^2} + \frac{y^2}{b^2} + \frac{z^2}{c^2} = 1$$

Hyperboloid of two sheets

$$\frac{x^2}{a^2} - \frac{y^2}{b^2} - \frac{z^2}{c^2} = 1$$

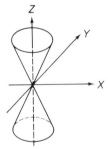

Elliptic cone

$$\frac{x^2}{a^2} + \frac{y^2}{b^2} - \frac{z^2}{c^2} = 0$$

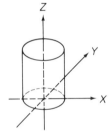

Elliptic cylinder

$$\frac{x^2}{a^2} + \frac{y^2}{b^2} = 1$$

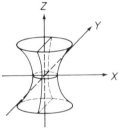

Hyperboloid of one sheet

$$\frac{x^2}{a^2} + \frac{y^2}{b^2} + \frac{z^2}{c^2} = 1$$

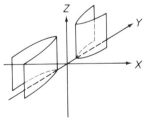

Hyperbolic cylinder

$$\frac{x^2}{a^2} - \frac{y^2}{b^2} = 1$$

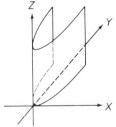

Parabolic cylinder

$$y^2 = px$$

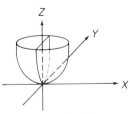

Elliptic paraboloid

$$\frac{x^2}{a^2} + \frac{y^2}{b^2} = 2cz$$

Figure 185

16.3. A QADRIC EXAMPLE: SHAPING A SWAGING DIE

This swaging die is used for hot-forging a section of a shaft or bar down to a smaller cross section. The die shape is a hyperbolic paraboloid (Fig. 184) with the equation

$$X^2 - Y^2 + 8Z - 16 = 0$$

The machining operation is finish-milling, using the paraboloid as part surface. The shape, therefore, is defined as

PTSURF = QADRIC/1,−1,0,0,0,0,0,0,8,−16

The cutter is a ball end mill of 0.500 diameter, with a corner radius of 0.25. CUTTER/0.5,0.25.

Since three-dimensional shapes are difficult to visualize, Fig. 186 gives a plot of the Z values at XY points over the surface of the finished die shape. The lines bordering the die, DIRGT, DIEND, and DILFT, are also indicated.

```
PARTNØ   SWAGING DIE SADDLE SECTION
CLPRNT
NØPØST
CUTTER/0.5,0.25
ØUTTØL/0.001
$$
SETPT  = PØINT/10,10,6
DIEND  = LINE/2,2,0,−2,2,0
DIRGT  = LINE/2,2,0,2,−2,0
DILFT  = LINE/−2,2,0,−2,−2,0
PTSURF = QADRIC/1,−1,0,0,0,0,0,0,8,−16
L1     = LINE/PARLEL,DIEND,YSMALL,.25
L2     = LINE/PARLEL,DIEND,YSMALL,.50
L3     = LINE/PARLEL,DIEND,YSMALL,.75
L4     = LINE/PARLEL,DIEND,YSMALL,1.0
L5     = LINE/PARLEL,DIEND,YSMALL,1.25
L6     = LINE/PARLEL,DIEND,YSMALL,1.50
```
and we continue to increment YSMALL by 0.25 until
```
L16    = LINE/PARLEL,DIEND,YSMALL,4.0
FRØM/SETPT
TLØNPS
GØ/ØN,DIEND,ØN,PTSURF,ØN,DIRGT
PSIS/PTSURF
```

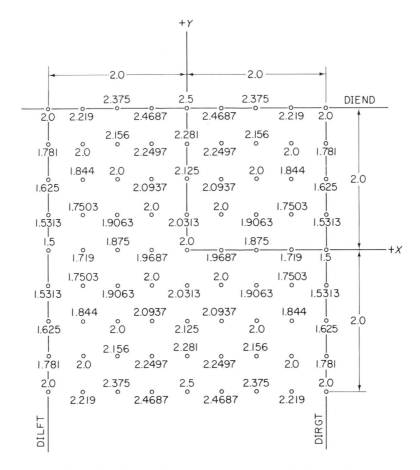

Fig. 186. Plot of Z values for a swaging die $X^2 - Y^2 + 8Z - 16 = 0$.

```
GØRGT/DIEND,ØN,DILFT
GØLFT/DILFT,ØN,L1
GØLFT/L1,ØN,DIRGT
GØRGT/DIRGT,ØN,L2
GØRGT/L2,ØN,DILFT
GØLFT/DILFT,ØN,L3
GØLFT/L3,ØN,DIRGT
GØRGT/DIRGT,ØN,L4
GØRGT/L4,ØN,DILFT
GØLFT/DILFT,ØN,L5
GØLFT/L5,ØN,DIRGT
GØRGT/DIRGT,ØN,L6
```

this sequence of movements continues to
GØRGT/L16,ØN,DILFT
GØDLTA/0,0,5
GØTØ/SETPT
FINI

There is a great deal of repetition in this program, both in the geometrical statements and the tool movement statements. The repetition can be eliminated by the use of a MACRØ and the familiar computer technique of the loop.

The small tolerance of 0.001 results in a printout of X positions about 0.25 in. apart. The full computer solution is produced as follows. Those who wish to sink this die shape should note that this is possible on any vertical-spindle mill that can position the two axes X and Z.

SWAGING DIE SADDLE SECTION

CUTTER/	.5000	.2500	
OUTTOL/	.0010	.0010	.0010
FROM/SETPT			
	X	Y	Z
	10.0000000	10.0000000	6.0000000
DS IS/DIEND			
	X	Y	Z
	2.0000000	2.0000000	2.0011023
DS IS/DIEND			
	X	Y	Z
	1.8234943	2.0000000	2.0854450
	1.6481850	2.0000000	2.1615080
	1.4739042	2.0000000	2.2295101
	1.3005445	2.0000000	2.2896207
	1.1279956	2.0000000	2.3419908
	.9561446	2.0000000	2.3867522
	.7848760	2.0000000	2.4240176
	.6140724	2.0000000	2.4538798
	.4436148	2.0000000	2.4764117
	.2733825	2.0000000	2.4916656
	.0836352	2.0000000	2.5001317
	−.1055736	2.0000000	2.4996129
	−.2943682	2.0000000	2.4901765
	−.4829266	2.0000000	2.4718594
	−.6714246	2.0000000	2.4446658
	−.8600373	2.0000000	2.4085663

	−1.0489387	2.0000000	2.3634991
	−1.2383005	2.0000000	2.3093702
	−1.4282909	2.0000000	2.2460541
	−1.6190748	2.0000000	2.1733944
	−1.7373204	2.0000000	2.1237942
	−1.8831451	2.0000000	2.0578123
	−2.0000000	2.0000000	2.0011023
DS IS/DILFT			
	X	Y	Z
	−2.0000000	1.8748736	1.9395151
	−2.0000000	1.7500000	1.8829326
DS IS/L1			
	X	Y	Z
	−1.8459435	1.7500000	1.9579406
	−1.6737082	1.7500000	2.0337020
	−1.5011417	1.7500000	2.1021728
	−1.3295358	1.7500000	2.1628811
	−1.1587795	1.7500000	2.2159824
	−.9887596	1.7500000	2.2616139
	−.8193600	1.7500000	2.2998930
	−.6504626	1.7500000	2.3309178
	−.4819473	1.7500000	2.3547665
	−.3136926	1.7500000	2.3714968
	.0610044	1.7500000	2.3833295
	.2476320	1.7500000	2.3761309
	.4339991	1.7500000	2.3602550
	.6202825	1.7500000	2.3357106
	.8066588	1.7500000	2.3024739
	.9933032	1.7500000	2.2604883
	1.1803883	1.7500000	2.2096653
	1.3680839	1.7500000	2.1498850
	1.5565557	1.7500000	2.0809969
	1.7459650	1.7500000	2.0028210
	1.8964920	1.7500000	1.9342982
	2.0000000	1.7500000	1.8838930
DS IS/DIRGT			
	X	Y	Z
	2.0000000	1.6267096	1.8308920
	2.0000000	1.5000000	1.7813679
DS IS/L2			
	X	Y	Z
	1.8468359	1.5000000	1.8559468
	1.6760635	1.5000000	1.9311338
	1.5051432	1.5000000	1.9990870

A Qadric Example: Shaping a Swaging Die

1.3352063	1.5000000	2.0594099
1.1661414	1.5000000	2.1122605
.9978346	1.5000000	2.1577777
.8301691	1.5000000	2.1960814
.6630259	1.5000000	2.2272721
.4962840	1.5000000	2.2514303
.3298206	1.5000000	2.2686162
.1441576	1.5000000	2.2796138
−.0409357	1.5000000	2.2820015
−.2255841	1.5000000	2.2758513
−.4099674	1.5000000	2.2612062
−.5942641	1.5000000	2.2380765
−.7786516	1.5000000	2.2064395
−.9633063	1.5000000	2.1662402
−1.1484017	1.5000000	2.1173918
−1.3341084	1.5000000	2.0597761
−1.5205928	1.5000000	1.9932448
−1.7080168	1.5000000	1.9176198
−1.8368238	1.5000000	1.8605560
−2.0000000	1.5000000	1.7823113

DS IS/DILFT

X	Y	Z
−2.0000000	1.3782967	1.7375797
−2.0000000	1.2500000	1.6954286

DS IS/L3

X	Y	Z
−1.8476182	1.2500000	1.7696314
−1.6781187	1.2500000	1.8443178
−1.5086288	1.2500000	1.9118194
−1.3401428	1.2500000	1.9718045
−1.1725490	1.2500000	2.0244326
−1.0057326	1.2500000	2.0698448
−.8395764	1.2500000	2.1081630
−.6739606	1.2500000	2.1394896
−.5087631	1.2500000	2.1639073
−.3438604	1.2500000	2.1814785
−.1599121	1.2500000	2.1930595
.0234593	1.2500000	2.1961865
.2063804	1.2500000	2.1909323
.3890318	1.2500000	2.1773411
.5715931	1.2500000	2.1554241
.7542431	1.2500000	2.1251601
.9371589	1.2500000	2.0864952
1.1205151	1.2500000	2.0393441

1.3044829	1.2500000	1.9835904
1.4892295	1.2500000	1.9190875
1.6749174	1.2500000	1.8456593
1.7901935	1.2500000	1.7957387
1.9416238	1.2500000	1.7251138
2.0000000	1.2500000	1.6963573

DS IS/DIRGT

X	Y	Z
2.0000000	1.1296188	1.6596202
2.0000000	1.0000000	1.6251146

DS IS/L4

X	Y	Z
1.8482779	1.0000000	1.6990002
1.6798451	1.0000000	1.7732667
1.5115528	1.0000000	1.8403890
1.3442819	1.0000000	1.9000890
1.1779203	1.0000000	1.9525277
1.0123530	1.0000000	1.9978479
.8474621	1.0000000	2.0361730
.6831270	1.0000000	2.0676075
.5192252	1.0000000	2.0922356
.3556322	1.0000000	2.1101218
.1731234	1.0000000	2.1221819
−.0088017	1.0000000	2.1259179
−.1902713	1.0000000	2.1214032
−.3714668	1.0000000	2.1086829
−.5525689	1.0000000	2.0877693
−.7337573	1.0000000	2.0586422
−.9152101	1.0000000	2.0212490
−1.0971026	1.0000000	1.9755056
−1.2796068	1.0000000	1.9212970
−1.4628906	1.0000000	1.8584781
−1.6471170	1.0000000	1.7868746
−1.7614560	1.0000000	1.7381679
−1.9336717	1.0000000	1.6586388
−2.0000000	1.0000000	1.6260312

DS IS/DILFT

X	Y	Z
−2.0000000	.8806627	1.5970598
−2.0000000	.7500000	1.5704259

DS IS/L5

X	Y	Z
−1.8488040	.7500000	1.6440587
−1.6812173	.7500000	1.7179918

A Qadric Example: Shaping a Swaging Die

−1.5138742	.7500000	1.7848126
−1.3475666	.7500000	1.8442851
−1.1821823	.7500000	1.8965716
−1.0176058	.7500000	1.9418161
−.8537188	.7500000	1.9801436
−.6904004	.7500000	2.0116596
−.5275272	.7500000	2.0364502
−.3649743	.7500000	2.0545810
−.1836091	.7500000	2.0670150
−.0028336	.7500000	2.0712271
.1774820	.7500000	2.0672915
.3575196	.7500000	2.0552541
.5374607	.7500000	2.0351281
.7174860	.7500000	2.0068940
.8977743	.7500000	1.9705003
1.0785017	.7500000	1.9258640
1.2598407	.7500000	1.8728712
1.4419597	.7500000	1.8113779
1.6250223	.7500000	1.7412112
1.8091866	.7500000	1.6621700
2.0000000	.7500000	1.5713328

DS IS/DIRGT

X	Y	Z
2.0000000	.6314090	1.5499476
2.0000000	.5000000	1.5313625

DS IS/L6

X	Y	Z
1.8491870	.5000000	1.6048113
1.6822138	.5000000	1.6785024
1.5155585	.5000000	1.7451043
1.3499492	.5000000	1.8044111
1.1852734	.5000000	1.8565862
1.0214155	.5000000	1.9017744
.8582567	.5000000	1.9401017
.6956758	.5000000	1.9716748
.5335492	.5000000	1.9965806
.3717511	.5000000	2.0148860
.1912163	.5000000	2.0275875
.0112756	.5000000	2.0321411
−.1682016	.5000000	2.0286213
−.3473978	.5000000	2.0170747
−.5264952	.5000000	1.9975150
−.7056750	.5000000	1.9699236
−.8851166	.5000000	1.9342496

	−1.0649967	.5000000	1.8904108
	−1.2454883	.5000000	1.8382941
	−1.4267601	.5000000	1.7777566
	−1.6089761	.5000000	1.7086260
	−1.7922950	.5000000	1.6307024
	−1.9422057	.5000000	1.5607364
	−2.0000000	.5000000	1.5322625
DS IS/DILFT			
	X	Y	Z
	−2.0000000	.3818591	1.5183393
	−2.0000000	.2500000	1.5079245
DS IS/L7			
	X	Y	Z
	−1.8494198	.2500000	1.5812614
	−1.6828184	.2500000	1.6548058
	−1.5165799	.2500000	1.7212748
	−1.3513938	.2500000	1.7804810
	−1.1871474	.2500000	1.8325881
	−1.0237250	.2500000	1.8777416
	−.8610078	.2500000	1.9160680
	−.6988741	.2500000	1.9476747
	−.5372002	.2500000	1.9726493
	−.3758600	.2500000	1.9910594
	−.1958289	.2200000	2.0039217
	−.0163949	.2100000	2.0086807
	.1625735	.2500000	2.0054112
	.3412590	.2500000	1.9941603
	.5198441	.2500000	1.9749421
	.6985105	.2500000	1.9477382
	.8774380	.2500000	1.9124984
	1.0568035	.2500000	1.8691409
	1.2367803	.2500000	1.8175530
	1.4175376	.2500000	1.7575924
	1.5992394	.2500000	1.6890675
	1.7820446	.2500000	1.6118390
	1.9393682	.2500000	1.5386706
	2.0000000	.2500000	1.5088203
DS IS/DIRGT			
	X	Y	Z
	2.0000000	.1320044	1.5022900
	2.0000000	0.0000000	1.5001118
DS IS/L8			
	X	Y	Z
	1.8494979	0.0000000	1.5734112

A Qadric Example: Shaping a Swaging Die

1.6830211	0.0000000	1.6469064
1.5169222	0.0000000	1.7133309
1.3518779	0.0000000	1.7725033
1.1877754	0.0000000	1.8245875
1.0244989	0.0000000	1.8697293
.8619296	0.0000000	1.9080553
.6999458	0.0000000	1.9396731
.5384236	0.0000000	1.9646706
.3772369	0.0000000	1.9831155
.1973747	0.0000000	1.9960315
.0181104	0.0000000	2.0008590
—.1606874	0.0000000	1.9976732
—.3392016	0.0000000	1.9865210
—.5176149	0.0000000	1.9674169
—.6961092	0.0000000	1.9403425
—.8748642	0.0000000	1.9052478
—1.0540571	0.0000000	1.8620512
—1.2338613	0.0000000	1.8106401
—1.4144460	0.0000000	1.7508724
—1.5959755	0.0000000	1.6825768
—1.7786082	0.0000000	1.6055540
—1.9384172	0.0000000	1.5313174
—2.0000000	0.0000000	1.5010062

DS IS/DILFT

X	Y	Z
—2.0000000	—.1181562	1.5018570
—2.0000000	—.2500000	1.5079245

DS IS/L9

X	Y	Z
—1.8494198	—.2500000	1.5812614
—1.6828184	—.2500000	1.6548058
—1.5165799	—.2500000	1.7212748
—1.3513938	—.2500000	1.7804810
—1.1871474	—.2500000	1.8325881
—1.0237250	—.2500000	1.8777416
—.8610078	—.2500000	1.9160680
—.6988741	—.2500000	1.9476747
—.5372001	—.2500000	1.9726493
—.3758600	—.2500000	1.9910594
—.1958289	—.2500000	2.0039217
—.0163949	—.2500000	2.0086807
.1625736	—.2500000	2.0054112
.3412590	—.2500000	1.9941603
.5198441	—.2500000	1.9749421

	.6985106	—.2500000	1.9477382
	.8774381	—.2500000	1.9124984
	1.0568035	—.2500000	1.8691409
	1.2367804	—.2500000	1.8175530
	1.4175376	—.2500000	1.7575924
	1.5992395	—.2500000	1.6890875
	1.7820446	—.2500000	1.6118390
	1.9393682	—.2500000	1.5386706
	2.0000000	—.2500000	1.5088203
DS IS/DIRGT			
	X	Y	Z
	2.0000000	—.3686199	1.5170973
	2.0000000	—.5000000	1.5313625
DS IS/L10			
	X	Y	Z
	1.8491870	—.5000000	1.6048113
	1.6822137	—.5000000	1.6785024
	1.5155585	—.5000000	1.7451043
	1.3499492	—.5000000	1.8044111
	1.1852734	—.5000000	1.8565863
	1.0214154	—.5000000	1.9017745
	.8582567	—.5000000	1.9401017
	.6956758	—.5000000	1.9716748
	.5335491	—.5000000	1.9965806
	.3717511	—.5000000	2.0148860
	.1912162	—.5000000	2.0275875
	.0112756	—.5000000	2.0321411
	—.1682016	—.5000000	2.0286213
	—.3473978	—.5000000	2.0170747
	—.5264952	—.5000000	1.9975150
	—.7056750	—.5000000	1.9699236
	—.8851167	—.5000000	1.9342496
	—1.0649968	—.5000000	1.8904108
	—1.2454883	—.5000000	1.8382941
	—1.4267601	—.5000000	1.7777565
	—1.6089762	—.5000000	1.7086260
	—1.7922950	—.5000000	1.6307024
	—1.9422057	—.5000000	1.5607364
	—2.0000000	—.5000000	1.5322625
DS IS/DILFT			
	X	Y	Z
	—2.0000000	—.6193801	1.5480669
	—2.0000000	—.7500000	1.5704259

A Qadric Example: Shaping a Swaging Die

DS IS/L11

X	Y	Z
−1.8488040	−.7500000	1.6440587
−1.6812172	−.7500000	1.7179918
−1.5138741	−.7500000	1.7848126
−1.3475665	−.7500000	1.8442851
−1.1821822	−.7500000	1.8965716
−1.0176057	−.7500000	1.9418162
−.8537188	−.7500000	1.9801436
−.6904003	−.7500000	2.0116597
−.5275272	−.7500000	2.0364502
−.3649742	−.7500000	2.0545810
−.1836091	−.7500000	2.0670150
−.0028336	−.7500000	2.0712271
.1774820	−.7500000	2.0672915
.3575196	−.7500000	2.0552541
.5374608	−.7500000	2.0351281
.7174861	−.7500000	2.0068940
.8977744	−.7500000	1.9705003
1.0785018	−.7500000	1.9258640
1.2598407	−.7500000	1.8728712
1.4419598	−.7500000	1.8113779
1.6250223	−.7500000	1.7412112
1.8091867	−.7500000	1.6621700
2.0000000	−.7500000	1.5713328

DS IS/DIRGT

X	Y	Z
2.0000000	−.8704271	1.5948193
2.0000000	−1.0000000	1.6251146

DS IS/L12

X	Y	Z
1.8482779	−1.0000000	1.6990002
1.6798450	−1.0000000	1.7732668
1.5115528	−1.0000000	1.8403890
1.3442818	−1.0000000	1.9000890
1.1779203	−1.0000000	1.9525277
1.0123529	−1.0000000	1.9978479
.8474620	−1.0000000	2.0361730
.6831269	−1.0000000	2.0676075
.5192251	−1.0000000	2.0922356
.3556321	−1.0000000	2.1101218
.1731233	−1.0000000	2.1221819
−.0088018	−1.0000000	2.1259179

−.1902714	−1.0000000	2.1214032
−.3714669	−1.0000000	2.1086829
−.5525690	−1.0000000	2.0877693
−.7337574	−1.0000000	2.0586422
−.9152101	−1.0000000	2.0212490
−1.0971026	−1.0000000	1.9755056
−1.2796069	−1.0000000	1.9212970
−1.4628907	−1.0000000	1.8584781
−1.6471171	−1.0000000	1.7868745
−1.7614560	−1.0000000	1.7381679
−1.9336717	−1.0000000	1.6586388
−2.0000000	−1.0000000	1.6260312

DS IS/DILFT

X	Y	Z
−2.0000000	−1.1217443	1.6574041
−2.0000000	−1.2500000	1.6954286

DS IS/L13

X	Y	Z
−1.8476181	−1.2500000	1.7696314
−1.6781186	−1.2500000	1.8443179
−1.5086287	−1.2500000	1.9118195
−1.3401428	−1.2500000	1.9718045
−1.1725489	−1.2500000	2.0244327
−1.0057325	−1.2500000	2.0698448
−.8395763	−1.2500000	2.1081630
−.6739605	−1.2500000	2.1394896
−.5087630	−1.2500000	2.1639073
−.3438604	−1.2500000	2.1814785
−.1599120	−1.2500000	2.1930595
.0234594	−1.2500000	2.1961865
.2063805	−1.2500000	2.1909323
.3890319	−1.2500000	2.1773411
.5715932	−1.2500000	2.1554241
.7542432	−1.2500000	2.1251601
.9371590	−1.2500000	2.0864952
1.1205151	−1.2500000	2.0393441
1.3044829	−1.2500000	1.9835904
1.4892296	−1.2500000	1.9190875
1.6749175	−1.2500000	1.8456592
1.7901936	−1.2500000	1.7957387
1.9416238	−1.2500000	1.7251137
2.0000000	−1.2500000	1.6963573

DS IS/DIRGT

2.0000000	−1.3733265	1.7358702

A Qadric Example: Shaping a Swaging Die

	2.0000000	−1.5000000	1.7813679
DS IS/L14			
	X	Y	Z
	1.8468358	−1.5000000	1.8559468
	1.6760634	−1.5000000	1.9311338
	1.5051431	−1.5000000	1.9990870
	1.3352062	−1.5000000	2.0594100
	1.1661414	−1.5000000	2.1122605
	.9978346	−1.5000000	2.1577777
	.8301691	−1.5000000	2.1960815
	.6630259	−1.5000000	2.2272722
	.4962839	−1.5000000	2.2514303
	.3298205	−1.5000000	2.2686162
	.1441575	−1.5000000	2.2796138
	−.0409358	−1.5000000	2.2820015
	−.2255842	−1.5000000	2.2758513
	−.4099675	−1.5000000	2.2612062
	−.5942641	−1.5000000	2.2380765
	−.7786517	−1.5000000	2.2064395
	−.9633063	−1.5000000	2.1662402
	−1.1484018	−1.5000000	2.1173918
	−1.3341084	−1.5000000	2.0597761
	−1.5205928	−1.5000000	1.9932447
	−1.7080168	−1.5000000	1.9176198
	−1.8368239	−1.5000000	1.8605560
	−2.0000000	−1.5000000	1.7823113
DS IS/DILFT			
	X	Y	Z
	−2.0000000	−1.6251493	1.8302577
	−2.0000000	−1.7500000	1.8829326
DS IS/L15			
	X	Y	Z
	−1.8459434	−1.7500000	1.9579406
	−1.6737081	−1.7500000	2.0337021
	−1.5011416	−1.7500000	2.1021729
	−1.3295357	−1.7500000	2.1628811
	−1.1587795	−1.7500000	2.2159825
	−.9887595	−1.7500000	2.2616139
	−.8193599	−1.7500000	2.2998930
	−.6504625	−1.7500000	2.3309178
	−.4819472	−1.7500000	2.3547665
	−.3136925	−1.7500000	2.3714968
	−.1260628	−1.7500000	2.3818085
	.0610044	−1.7500000	2.3833295

	.2476321	−1.7500000	2.3761309
	.4339991	−1.7500000	2.3602550
	.6202826	−1.7500000	2.3357106
	.8066589	−1.7500000	2.3024739
	.9933032	−1.7500000	2.2604883
	1.1803883	−1.7500000	2.2096653
	1.3680839	−1.7500000	2.1498850
	1.5565558	−1.7500000	2.0809969
	1.7459651	−1.7500000	2.0028210
	1.8964921	−1.7500000	1.9342981
	2.0000000	−1.7500000	1.8838930
DS IS/DIRGT			
	X	Y	Z
	2.0000000	−1.8771956	1.9406042
	2.0000000	−2.0000000	2.0001225
DS IS/L16			
	X	Y	Z
	1.8449540	−2.0000000	2.0756065
	1.6710829	−2.0000000	2.1520092
	1.4966730	−2.0000000	2.2210571
	1.3231989	−2.0000000	2.2821922
	1.1505502	−2.0000000	2.3355681
	.9786143	−2.0000000	2.3813191
	.8072761	−2.0000000	2.4195604
	.6364184	−2.0000000	2.4503876
	.4659223	−2.0000000	2.4738760
	.2956674	−2.0000000	2.4900808
	.1058431	−2.0000000	2.4996058
	−.0834253	−2.0000000	2.5001361
	−.2722591	−2.0000000	2.4917421
	−.4608361	−2.0000000	2.4744649
	−.6493320	−2.0000000	2.4483124
	−.8379223	−2.0000000	2.4132592
	−1.0267810	−2.0000000	2.3692472
	−1.2160800	−2.0000000	2.3161861
	−1.4059882	−2.0000000	2.2539541
	−1.5966707	−2.0000000	2.1823985
	−1.7882881	−2.0000000	2.1013366
	−1.9410963	−2.0000000	2.0301149
	−2.0000000	−2.0000000	2.0011023
DS IS			
	X	Y	Z
	−2.0000000	−2.0000000	7.0011023

```
DS IS/SETPT
              X              Y              Z
         10.0000000     10.0000000      6.0000000

FINI
END OF PART PROGRAM
```

Sometimes what is desired is a mathematical analysis of, or a printout of points for, such a quadratic surface as this. This is obtained by the use of a CUTTER/0 statement.

16.4. TABCYL

TABCYL is an abbreviation of "tabulated cylinder," which is a smooth cylindrical surface passing through an arbitrary set of points. The points may have no apparent mathematical relationship to one another. The word "cylinder" here is used in the mathematical meaning of a general cylinder and not a circular cylinder (which in APT is a CYLNDR), as in Fig. 183. The TABCYL extends to infinity in both directions from the given points. The term "tabulated" simply means that the surface is defined by a series of points that can be tabulated in a table of coordinates.

The TABCYL routine is of very great usefulness when a surface must be defined by an empirical series of points obtained from measured data or other unmathematical means. There is a variety of possible statements in APT language for defining a TABCYL, but only one of these is discussed here, and it is restricted to two dimensions. This is the format to be explained:

TABULC = TABCYL/NØZ,SPLINE,PT1,PT2,PT3, - - - PTn.

Either NØX (no X values), NØY (no Y values), or NØZ (no Z values) may be used. For the NØZ case, the TABCYL is perpendicular to the XY plane, and only XY values are calculated. The points defining the TABCYL, of which there must be at least four, may be symbolic names or XY values (or ZX or YZ, in this order of coordinates) and must be in proper sequence along the TABCYL.

The use of a TABCYL statement to define a closed or a self-intersecting curve may be unsuccessful. If the curve through the points should be intersecting, it may be necessary to divide it into two TABCYL's with three overlapping points. However, the second TABCYL cannot be used as a check surface for the other TABCYL as drive surface. These restrictions will be further explained.

16.5. A TABCYL APPLICATION: THE SCROLL CASING FOR A CENTRIFUGAL FAN

Back in Chapter 7, the problem of fitting a curve to the scroll casing of a centrifugal fan by mathematical methods was examined—and hastily dropped, since it would call for prodigies of computation. The problem of the fan scroll would not be complex in terms of a computer and a suitable computing language, however.

We can solve the problem of cutting or machining or drafting a fan scroll by the use of the TABCYL routine. The proportions of the scroll casing are defined in terms of fan diameter D in the first drawing of part print no. 5. Drawing no. 2 is dimensioned for a wheel diameter of 20 in., with symbolic names for the geometric elements. (No reader should assume that all fan manufacturers conform exactly to these proportions in their designs.)

In order to program this shape in APT language, we must define the small circular fillet connecting the TABCYL and LINB. Here we will define the fillet by the XY coordinates of its center and its radius.

Let us set up the geometric definitions as follows:

```
LINB   = LINE/20,26.459,40,26.459
LINF   = LINE/33.420,41.240,33.420,0
LINT   = LINE/20,41.24,40,41.24
CIRCF  = CIRCLE/31.687,25.695,0.76
TABFAN = TABCYL/NØZ,SPLINE,31.004,25.366,32.920,20 $
         30.110,9.890,20,4.320,7.939,7.939,1.560,20 $
         5.974,34.026,20,41.240
```

CIRCF and LINT are presumed to be tangent to TABFAN, and probably are, at least to three decimal places. We shall program the tool from LINB to CIRCF to TABFAN to LINT. TABFAN will be the check surface at the end of CIRCF and LINT will be the check surface after programming around TABFAN. Unfortunately, we have here that rather familiar misfit, the "perfect" APT program that the computer cannot solve. It is possible that this program could succeed, especially with generous tolerances, but it is a type of program that is not usually successful. Stopping a tool movement TANTØ a TABCYL is not recommended practice; also, the tangency of LINT is assumed and not certain.

To improve this program, we first require a circle definition that has not so far appeared in this book. This new definition is for a circle that is (1) tangent to a line, (2) tangent to a TABCYL, and (3) of a given radius. The format is

A Tabcyl Application

$$\text{CIRCF} = \text{CIRCLE/TANT}\emptyset,\text{(line)},\begin{matrix}\text{XLARGE}\\\text{XSMALL}\\\text{YLARGE},\\\text{YSMALL}\end{matrix}\text{(tabcyl)},\begin{matrix}\text{XLARGE}\\\text{XSMALL}\\\text{YLARGE},\\\text{YSMALL}\end{matrix}\text{(nearest}$$
point on the TABCYL),RADIUS,(r)

The first XYLARGE-SMALL modifier is used to exclude circles lying on the other side of the line. For CIRCF use YSMALL. The second XYLARGE-SMALL modifier specifies on which side of the TABCYL the circle lies. There are cases where no second ambiguity exists and the second XYLARGE-SMALL modifier could be omitted, but the format requires that this second modifier be included, even if it is an arbitrary choice. The point on the TABCYL should be the defined point that lies nearest the point of tangency; it is only rarely the point of tangency, as in this case.

The fillet circle is thus defined:

CIRCF = CIRCLE/TANT∅,LINB,YSMALL,TABFAN,XLARGE,(P∅INT/31.004, $
25.366,0),RADIUS,0.76

Note an important difference. With this definition we are requiring the computer to satisfy itself that this circle is tangent to the TABFAN. In the previous attempt, the circle was defined by its center and radius; tangency was assumed. With only three decimal places, tangency may not occur to the satisfaction of the computer.

The above attempt to program the fan scroll had some errors that are apparent only with experience in APT. It will be instructive next to introduce a deliberate error into the program, because the computer printout then discloses something of what a computer does with a TABCYL. Suppose we define the fillet circle as

CIRCF = CIRCLE/TANT∅,LINB,YLARGE,TABFAN,XLARGE,(P∅INT/31.004 $
25.366,0),RADIUS,0.76

We have changed the YSMALL to YLARGE, thus defining a circle tangent to LINB and TABFAN, but this circle now lies above LINB instead of below. If we program the tool from LINB to CIRCF to TABFAN, we will run the tool *off* the TABFAN upwards. Here is the printout for this erroneous program:

FR∅M/SETPT

	X	Y	Z
	4.0000000	50.0000000	0.0000000
DS IS/LINT			
	33.4200000	41.2400000	0.0000000
DS IS/LINF			
	33.4200000	26.4590000	

DS IS/LINB
    30.7660395   26.4590000

DS IS/CIRCF
    30.6234859   26.4708603
    30.4839592   26.5115643
    30.3478128   26.5825088
    30.2733375   26.6382425
    30.2019339   26.7073170
    30.1154343   26.8230865

DS IS/TABFAN
    24.9169903   35.3656962
    24.7974614   35.5682906
    24.5529209   35.9701443
    24.0638400   36.7738519
    and the printout continues to
    21.3948800   41.1597569

The XY values from (30.1154343,26.8230865) to (21.3948800, 41.1597569) lie on a straight line. The computer, therefore, extends a tangent at each end of a TABCYL. The computer, of course, is programmed to cut this tangent off at a fixed length. What we have done in this program is to send the computer off in the wrong direction looking for the next check surface, which is presumably LINT, the next surface after TABFAN. The tangent extended to $Y = 41.1597569$, almost to LINT, which is located at $Y = 41.240$.

We can now program the fan scroll casing in a proper manner. There is, however, another potential hazard to be avoided. The tool must move around TABFAN and on to LINT. If we were to use LINT as a check surface thus,

GØFWD/TABFAN,TANTØ,LINT

we would almost certainly be in trouble again. Check surfaces TANTØ TABCYL's are not successful, but TØ,ØN, or PAST for the check surface will succeed. So we define a vertical line to pass through the point of tangency of TABFAN and LINT and use it for a check surface. This line is

LINC = LINE/20,41.240,20,0

The following is a successful program.

```
PARTNØ   FAN SCROLL 20WHEEL
CUTTER/0
CLPRNT
```

```
INTØL/.002
ØUTTØL/.002
NØPØST
LINB    = LINE/20,26.459,40,26.459
LINF    = LINE/33.42,41.24,33.42,0
LINT    = LINE/20,41.24,40,41.24
TABFAN  = TABCYL/NØZ,SPLINE,31.004,25.366,32.920,20,30.110,9.890,$
          20,4.320,7.939,7.939,1.560,20,5.974,34.026,20,41.240
CIRCF   = CIRCLE/TANTØ,LINB,YSMALL,TABFAN,XLARGE,$
          (PØINT/31.004,25.366,0),RADIUS,0.76
LINC    = LINE/20,41.240,20,0
SETPT   = PØINT/40,50,0
PT1     = PØINT/30,0,0
PARTS   = PLANE/SETPT,PT1,(PØINT/0,0,0)
$$ TOOL MOVEMENTS FOLLOW
FRØM/SETPT
GØ/ØN,LINT,ØN,PARTS,ØN,LINF
PSIS/PARTS
GØLFT/LINF,ØN,LINB
GØRGT/LINB,TANTØ,CIRCF
GØFWD/CIRCF,TANTØ,TABFAN
GØFWD/TABFAN,ØN,2,INTØF,LINC
GØFWD/LINT,ØN,LINF
GØTØ/SETPT
FINI
```

The following tabulation reproduces the complete CLPRNT for CIRCF and TABFAN as drive surfaces.

DS IS/CIRCF

X	Y	X	Y
31.5482631	26.4471397	30.9417646	25.8366788
31.4087364	26.4064357	30.9293712	25.6836580
31.2725900	26.3354912	30.9369211	25.5909444
31.1534378	26.2386836	30.9557174	25.4998428
31.0561212	26.1199467	30.9931089	25.3936504
30.9845947	25.9841052	31.0402116	25.3030865

DS IS/TABFAN

X	Y	X	Y
31.1745130	25.0816393	22.7137833	4.8409369
31.3895526	24.6959471	22.0449457	4.6671338

31.6640642	24.1581059	21.3561566	4.5203783
31.9137009	23.6124232	20.9202327	4.4433508
32.1631742	22.9939635	20.2587159	4.3485443
32.3817032	22.3666288	20.0000000	4.3183904
32.5697335	21.7298993	19.6673140	4.2851225
32.7276269	21.0830645	19.0006122	4.2372559
32.8556014	20.4253663	18.3393002	4.2151827
32.9537333	19.7560063	17.6876902	4.2189378
33.0149661	19.1545055	17.0451706	4.2481907
33.0497424	18.5621290	16.4111386	4.3026948
33.0585144	17.9781484	15.7850031	4.3822891
33.0416250	17.4018872	15.0935287	4.5009115
32.9993113	16.8326796	14.4125191	4.6503269
32.9317028	16.2698724	13.7413551	4.8303472
32.8264469	15.6487567	13.0792283	5.0409556
32.6902023	15.0363680	12.4253810	5.2822667
32.5230805	14.4319337	11.7790727	5.5545365
32.3249980	13.8345614	11.1396153	5.8581491
32.0956999	13.2433976	10.5063630	6.1936182
31.8347649	12.6576346	9.8787168	6.5615820
31.5416089	12.0765129	9.2561264	6.9627993
31.2154906	11.4993241	8.6380948	7.3981426
30.8555170	10.9254139	8.0241667	7.8686028
30.4606504	10.3541852	7.4142561	8.3755866
30.0297009	9.7851003	6.8968051	8.8406528
29.6188383	9.2863508	6.4054968	9.3164352
29.1967090	8.8138349	5.9396296	9.8030755
28.7631639	8.3667213	5.4985581	10.3007674
28.3180236	7.9442828	5.0816990	10.8097529
27.8610494	7.5458632	4.6885333	11.3303214
27.3919440	7.1708808	4.3186123	11.8628045
26.9103592	6.8188356	3.9328590	12.4714258
26.3577462	6.4525649	3.5761474	13.0940869
25.7904588	6.1149972	3.2479814	13.7311694
25.2080767	5.8055725	2.9478894	14.3832537
24.6099669	5.5237720	2.6755126	15.0509875
23.9954137	5.2692090	2.4306096	15.7350727
23.3636357	5.0416354	2.2130538	16.4362678

CONT./TABFAN

X	Y	X	Y
2.0228333	17.1553814	6.1497117	34.2407061
1.8600498	17.8932689	6.6493177	34.8169716

1.7249180	18.6508195	7.1584248	35.3654204
1.6177583	19.4289850	7.6771244	35.8868042
1.5389958	20.2287329	8.2055457	36.3818197
1.4936557	20.9608925	8.7438579	36.8511103
1.4745018	21.6823673	9.2922701	37.2952633
1.4811456	22.3938188	9.8510310	37.7148089
1.5132745	23.0958181	10.4204279	38.1102180
1.5706472	23.7889268	11.0007861	38.4819009
1.6530941	24.4736956	11.6616894	38.8690173
1.7605184	25.1506636	12.3354910	39.2265355
1.9098876	25.8976746	13.0225453	39.5549016
2.0899043	26.6343960	13.7234016	39.8545463
2.3004625	27.3614118	14.4386611	40.1257976
2.5416026	28.0794499	15.1689733	40.3688801
2.8134783	28.7892038	15.9150328	40.5839151
3.1163589	29.4913421	16.6775761	40.7709210
3.4506221	30.1864902	17.4573775	40.9298138
3.8167547	30.8752379	18.2552467	41.0604090
4.2153485	31.5581351	19.0720248	41.1624231
4.6470978	32.2356902	19.7396150	41.2232011
5.1127921	32.9083637	20.0000000	41.2416034
5.6133268	33.5765882		

16.6. ANALYZING THE TABCYL

A computer routine that can provide a faired curve through a series of points is a most useful one. The TABCYL routine, unfortunately, is not quite that versatile. It will provide a curve through a reasonably smooth set of points, but may not be able to approximate a curve that has radical changes of curvature. Sometimes an erratic curve can be made acceptable by increasing the number of points used. The TABCYL example of the previous section used a smooth spiral curve, where the length of the radius of curvature is proportional to the angle turned by the curve, and thus was certain to be acceptable to the computer.

Sometimes the computer cannot find an approximation that will fit the set of points. When this occurs, one or more points may have to be adjusted in coordinates. To assist in these adjustments, the computer prints out considerable information about the curve, and in this section an abbreviated discussion of this printout will be given. It must be noted that there may be small differences in this printout from one computer to another.

For every TABCYL the computer offers a curvature plot. This plot for the TABFAN is shown in Fig. 187. The curvatures (curvature =

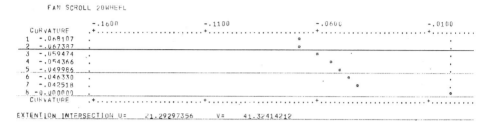

```
FAN SCROLL 20WHEEL
                -.1600              -.1100              -.0600              -.0100
   CURVATURE   .+..................+...................+...................+..........
 1  -.068107    .                   .                   .    o              .
 2  -.067387    .                   .                   .   o               .
 3  -.059474    .                   .                   .          o        .
 4  -.054366    .                   .                   .              o    .
 5  -.049986    .                   .                   .                  o
 6  -.046330    .                   .                   .                   .   o
 7  -.042518    .                   .                   .                   .       o
 8  -0.000000   .                   .                   .                   .                   o
   CURVATURE   .+..................+...................+...................+..........
EXTENTION INTERSECTION U=    21.29297356     V=     41.32414212
```

Fig. 187. Curvature plot for TABFAN.

reciprocal of the radius) are measured at each point supplied in the TABCYL definition statement. For the TABFAN, this plot is a smooth one of decreasing curvatures (increasing radius), and all curvatures are minus (concave). If the sign of the curvatures changes from plus to minus, then this sign change indicates an S curve or reverse curvature, which may or may not be indicative of difficulty in curve-fitting by the computer.

In addition to this curvature plot, the computer prints out analytical tabulations. The CDC 6600 computer used for the TABFAN program provides three such tabulations, shown for the TABFAN in Figs. 188, 189, and 190. The tabulated data of Fig. 188 include the actual X and Y co-

NUM	THETA	RADIUS	X-CORD	Y-CORD	SEG LENGTH	SEG ANGLE	EXT ANGLE
1	39.2684	40.058482	31.004000	25.366000	5.697808	-70.3502	.1540
2	31.2801	38.519169	32.920000	20.000000	10.493245	74.4671	-35.1826
3	19.1834	31.692652	30.110000	9.690000	11.542833	28.8521	-45.6150
4	12.1886	20.361241	20.000000	4.320000	12.592255	-16.7023	134.4456
5	45.0000	11.227441	7.959000	7.939000	13.644023	-62.1259	-45.4236
6	85.5400	20.060748	1.560000	20.000000	14.704152	72.5311	-45.3430
7	80.0420	34.546452	5.974000	34.026000	15.772459	27.2181	-45.3130
8	64.1282	45.833804	20.000000	41.240000	0.000000	0.0000	0.0000

Figure 188

ordinates of the defined points in the TABCYL statement, segment length, and segment angle. The segment length is the straight-line distance from the designated point to the next point in the TABCYL definition. Figure 191 shows the segment length between the first two points of the TABFAN definition.

The tabulation of Fig. 189 gives seven data items for each point in the TABCYL definition. These will not be discussed, except to note that the values of "delta curvature" should preferably be 0.001 or less. If the value of delta curvature should be too high, then the coefficients of the cubic equations that define the TABCYL are possibly unacceptable. If so,

NUM	SLOPE	NORMAL	ALFHA	TANGENT A	TANGENT B	CURV A	DELTA CURV
1	-1.64330	31.3218	-7.9066	.2065813	-.2065813	-.0661	0.0000
2	-7.13551	7.9777	-23.3024	.4350323	-.4115337	-.0674	.0007
3	1.28448	-37.9017	-56.0851	.4295552	-.4166044	-.0595	-.0010
4	.10926	-83.7647	-95.9533	.4231898	-.4081277	-.0544	.0006
5	-.80701	51.0960	6.0960	.4290519	-.4187302	-.0500	-.0004
6	-11.08763	5.1536	-80.3864	.4167197	-.4003403	-.0463	.0007
7	1.22232	-39.2871	-119.3291	.4347042	-.4347042	-.0425	-.0007
8	.06508	-86.2766	-150.4048	0.0000000	0.0000000	-0.0000	0.0000

Figure 189

the point on the TABCYL that requires adjustment may be the point with the high delta curvature or a point preceding it.

The method used by the computer to provide the TABCYL curve is explained with the assistance of Fig. 190.

U	V	A	B	LENGTH	MAX	MIN
21.39008383	41.16456070	0.00000000	0.00000000	18.49383423	0.00000000	0.00000000
31.00400000	25.36600000	0.00000000	-.03625629	5.69780765	.05164534	0.00000000
32.92000000	20.00000000	.00021341	-.04369772	10.49324545	.10584112	0.00000000
30.11000000	9.89000000	.00009720	-.03833599	11.54283327	.10577613	0.00000000
20.00000000	4.32000000	.00009499	-.03480329	12.59225484	.10392321	0.00000000
7.93900000	7.93900000	.00005545	-.03220264	13.64402294	.10597669	0.00000000
1.56000000	20.00000000	.00007576	-.02945420	14.70415152	.10214276	0.00000000
5.97400000	34.02600000	0.00000000	-.02756096	15.77245929	.10867605	0.00000000
20.00000000	41.24000000	0.00000000	0.00000000	1.28275142	0.00000000	0.00000000
21.28004383	41.32330070					

Fig. 190. Data for cubic equations of the TABCYL.

TABFAN was a NØZ or XY curve. The computer, however, calculates the curve not in the programmer's XY coordinate system but a rotated U'V' coordinate system. There is a U'V' coordinate system for each point in the TABCYL definition, the origin of coordinates in every case being the given TABCYL point. The U' axis passes through the following point, and the V' axis is, of course, perpendicular to the U' axis. These relationships are illustrated in Fig. 191. Each two successive points define a TABCYL interval with its U'V' coordinate system.

The equation of the TABCYL in each interval is a cubic equation of the form $Au^3 + Bu^2 + Cu + D = v$. But since the first point of the interval is used as origin, the D term becomes zero. The TABCYL is, therefore, a series of cubic curves of the form $Au^3 + Bu^2 + Cu = v$.

In Fig. 190, notice that the first two columns, U and V, are the X and Y coordinates of the TABCYL points. The next two columns, A and B, are coefficients for the cubic equation in the given interval, $Au^3 + Bu^2 + Cu$. The coefficient C is not printed out in most computer systems. C is calculated from

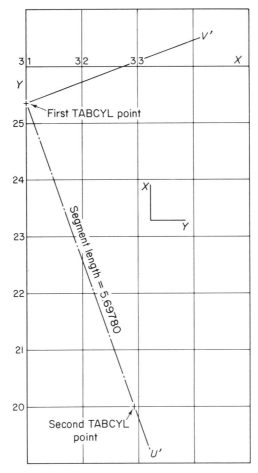

. **Figure 191**

$$C = -(AL^2 + BL)$$

where L is the segment length given in the fifth column of Fig. 190.

The printout of Fig. 190 has 10 points. The first and last of the 10 points are the ends of the tangent extensions of the TABCYL; the other eight points are the points given in the TABCYL definition. The second row in the printout, therefore, refers to the first defined point of TABFAN, 31.004, 25.366. For the curve between the first and second points of TABFAN

$$A = 0, \quad B = -0.036256$$

and $C = -(-0.036256 \times 5.6978) = +0.20658$. The equation of the TABFAN curve in U'V' coordinates between the first two points is

$$-0.036256u^2 + 0.20658u = v$$

U	V($= -0.036256U^2 + 0.20658U$)
1.0	+0.17033
2.0	+0.26814
3.0	+0.29364
4.0	+0.24622
5.0	+0.12650
5.6978	0.0

These values of U and V are plotted in Fig. 192.

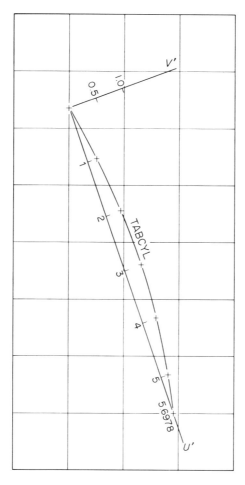

Figure 192

For the TABFAN curve between the second and third points, from Fig. 190

$$A = 0.00021341$$
$$B = -0.043698$$
$$C = -(0.00021341 \times 10.49325^2 - 0.043698 \times 10.49325)$$
$$= +0.435036$$

and the equation of the curve in U'V' coordinates is

$$0.00021341U^3 - 0.043698U^2 + 0.435036U = V$$

U	V	U	V
1.0	+0.39155	7.0	+0.977249
2.0	+0.696987	8.0	+0.792881
3.0	+0.917588	9.0	+0.531361
4.0	+1.054634	10.0	+0.19390
5.0	+1.109406	10.49325	0.0
6.0	+1.083184		

The last two columns in Fig. 190 are the maximum and minimum values of V expressed as a fraction of the segment length given in the fifth column.

Questions

1. Design a centrifugal fan scroll casing of a size capable of being machined by your n/c machine.
2. Program the contouring of this casing, choosing suitable tolerances. Machine it in heavy-gauge steel or aluminum sheet.
3. For the same casing, determine the constants A, B, and C for the cubic equations in U'V' for the first and second segments of the TABCYL. Plot U and V points graphically in the manner demonstrated in the text.

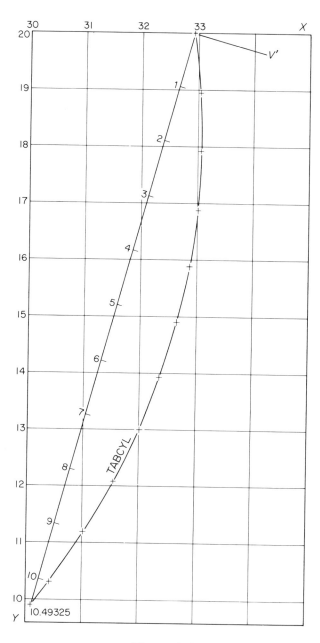

Figure 193

17

CREATIVITY IN NUMERICAL CONTROL

Some of the limitations of numerical control technology have been suggested throughout this book. In its present state, numerical control is an imperfect information science in many respects. The coordination between the information, the information processing, the computer, and the information-controlled machine is imperfect. Automatic input of information is almost unknown. No continent-wide numerical-control network has been developed, even though the general technology and hardware for the purpose is available. Even the mundane matter of automatically processing a series of rough and finishing cuts to reach a final finished shape has been given little attention (the simple case of the APT PØCKET/ routine of Chapter 14 is an exception to this statement); rather, numerical-control procedures emphasize the final cut that produces the final shape.

Despite these and other limitations, numerical-control technology represents a dazzling array of quite remarkable techniques both in machines and in information processing. The APT system has its limitations, yet it is an outstanding achievement. Since we are supplied with such a fund of technical resources, the most serious limitation is the creativity and imagination of the personnel who use numerical-control methods. Therefore, in this final chapter it is stimulating to explore the frontiers of this engineering science.

17.1. A TECHNICAL FRONTIER: ARTIFICIAL LIMBS

An engineering area that has risen rapidly to prominence is that of bioengineering. The practice of bioengineering embraces both medicine and engineering, with special emphasis on materials engineering. This is stimulating work, and as a result there have been many remarkable achievements. The manufacture and use of artificial organs and parts for the human body have attracted wide interest.

Is it possible to manufacture artificial limbs by using the methods of numerical control? Since this would seem to be a subject worthy of investigation, let us examine these matters here.

An artificial limb, a prosthetic device, is made to match the surviving limb. In order to make such a limb, much information about the shape of the limb must be processed. On first acquaintance, this appears to be an excellent application for numerical control.

The science and art of prosthetics, like numerical control, is highly sophisticated. Artificial limbs, in the present state of the art, are made by methods depending heavily on skills of hand and eye. In the following pages, the problem of an artificial limb will be discussed in terms of a BK model leg (BK—below the knee, AK—above the knee). A shank or a stump model is either sculptured or cast in a plaster impression of the surviving limb. The BK model must be fitted to the stump at the knee and must have a foot at the other end. Rigid foamed polyurethane with a density of 12 lb/cu ft is commonly used. There are many operations to the sequence of making and fitting the limb, including the cosmetic problem of simulating the appearance of a real limb. The complexity of the operations and materials is such as to make any suitable discussion hardly possible here.

To use numerical-control methods, we must first be able to collect the required data from the surviving limb. This is a more or less routine problem in instrumentation. The best method of obtaining the data automatically would appear to be that of traversing the limb contours with an ultrasonic proximity sensor and reading out dimensions as electrical signals. The data could be automatically punched into tape or cards and these mailed to a center that manufactures the limbs. Or the data could be transmitted by any available long-range method.

The institutional center would use a library of standard punched card programs for limb-shaping. On receipt of the data for a specific limb, these data, if not already punched, would be punched into cards which would be added to the standard deck. A standard BK computer routine then would provide the punched tape.

Limb shaping begins with a suitable preform, of which there must be more than one size for each type of limb. There are a number of preform considerations which perhaps are not entirely pertinent to a discussion of numerical-control methods. The preform, after it is shaped, must, for example, have a socket for receipt of the stump. The socket lining must be a deformable material such as flexible foam or rubber, so that it will mold to the shape of the stump.

After the preform is machined to shape, it must be ground and polished smooth and the required cosmetic treatments applied. It is then sent away for fitting to the patient.

17.2. SCULPTURING A BK MODEL

A BK model limb can be produced on any two-axis positioning mill, and a method for doing so is presented here.

Figure 194 presents a tabulation of dimensions for a BK amputation. Data were obtained by means of a female cast and a male cast, measurements being made on the male cast. As the tabulation suggests, the measurements were made by rotation of the cast by 15-deg increments with measurements every inch along the cast to its length of 15 in. The measurements are given to greater accuracy than is needed for an artificial limb. Perhaps a precision of 0.01 in. is more than sufficient.

To interpret these dimensions, they must be read as distances from an unknown center line of the limb. This need not be confusing. Consider a cross section of the limb at zero inches. The dimensions for zero inches are simply laid out radially from a center point. This has been done in Fig. 195. Before undertaking the sculpture of such a limb, a few sectional and longitudinal plots of the data should be made, in order to anticipate any difficulties and to understand the orientation of the cutter to the finished shape.

Making the preform presents few problems. Find or make a polyethylene bag of suitable length and diameter. Fill the bag with polyurethane foam. A convenient method of foaming the preform is the use of a Froth-Pak container. This contains the two components of the polyurethane mix with a suitable applicator. Froth-Paks are sold through local distributors, or information may be obtained from the manufacturer, Insta-Foam Products, Inc., 880 South Fiene Drive, Addison, Illinois, 60101. Froth-Pak kits supply a foam density of 2.7 lb/cu. ft. This is, of course, not the proper density, but is cheaper and entirely suitable for an investigational model limb. A polyethylene bag is suggested because polyurethane is extremely adhesive to most materials, but can be released from polyethylene, Teflon, or silicone rubber. If a more permanent mold is desired, it can be made of any convenient material with a finish coat of silicone rubber applied with a paint brush.

(If the user has no previous experience with a Froth-Pak kit, a warning is needed. The release valve on the container must be pulled all the way out. The result of incomplete opening of this valve is a filthy mess and a false conclusion that the material is worthless.)

Degrees / Inches	0°	15°	30°	45°	60°	75°	90°	105°	120°	135°	150°	165°	180°	195°	210°	225°	240°	255°	270°	285°	300°	315°	330°	345°
0 inch	1.580	1.485	1.325	1.215	1.180	1.220	1.335	1.440	1.500	1.575	1.750	1.985	2.160	2.110	1.910	1.730	1.605	1.460	1.295	1.185	Min 1.160	1.210	1.320	1.495
1	1.432	1.390	1.290	1.205	1.170	1.190	1.245	1.310	1.385	1.465	1.585	1.745	1.880	1.890	1.780	1.645	1.520	1.410	1.295	1.215	1.175	1.190	1.265	1.370
2	1.380	1.370	1.310	1.240	1.210	1.215	1.250	1.300	1.360	1.440	1.545	1.665	1.770	1.760	1.675	1.565	1.485	1.410	1.330	1.265	1.225	1.235	1.280	1.340
3	1.420	1.410	1.380	1.340	1.315	1.315	1.340	1.370	1.425	1.495	1.575	1.675	1.750	1.740	1.680	1.610	1.550	1.500	1.445	1.395	1.355	1.350	1.370	1.400
4	1.550	1.530	1.520	1.500	1.475	1.460	1.470	1.490	1.530	1.590	1.650	1.740	1.805	1.790	1.760	1.730	1.700	1.675	1.650	1.610	1.565	1.545	1.550	1.555
5	1.745	1.715	1.715	1.700	1.670	1.650	1.640	1.640	1.670	1.705	1.755	1.840	1.895	1.875	1.870	1.880	1.890	1.900	1.890	1.865	1.825	1.800	1.795	1.770
6	1.950	1.915	1.920	1.910	1.885	1.850	1.810	1.790	1.790	1.815	1.865	1.970	2.010	1.975	1.975	2.000	2.055	2.100	2.110	2.110	2.080	2.150	2.030	1.995
7	2.145	2.115	2.125	2.125	2.090	2.145	1.985	1.925	1.895	1.900	1.975	2.080	2.125	2.080	2.060	2.100	2.160	2.020	2.260	2.285	2.270	2.250	2.220	2.185
8	2.280	2.265	2.305	2.310	2.275	2.220	2.140	2.045	1.985	1.970	2.030	2.145	2.185	2.130	2.100	2.140	2.195	2.265	2.335	2.390	2.405	2.395	2.350	2.305
9	2.360	2.365	2.425	2.435	2.410	2.345	2.235	2.115	2.035	2.000	2.050	2.155	2.190	2.125	2.100	2.130	2.180	2.240	2.325	2.410	2.440	Max 2.450	2.410	2.370
10	2.365	2.380	2.470	2.485	2.470	2.410	2.295	2.165	2.080	2.040	2.065	2.150	2.180	2.115	2.060	2.085	2.130	2.170	2.235	2.325	2.390	2.410	2.385	2.365
11	2.325	2.345	2.415	2.445	2.450	2.410	2.315	2.190	2.100	2.050	2.065	2.160	2.185	2.115	2.055	2.050	2.065	2.080	2.130	2.190	2.265	2.290	2.300	2.310
12	2.190	2.205	2.265	2.310	2.360	2.355	2.295	2.185	2.090	2.055	2.085	2.190	2.215	2.140	2.080	2.040	2.020	2.020	2.040	2.100	2.150	2.160	2.175	2.190
13	2.130	2.045	2.100	2.175	2.265	2.300	2.250	2.170	2.080	2.065	2.130	2.240	2.265	2.190	2.115	2.060	2.010	1.985	2.000	2.075	2.110	2.110	2.080	2.055
14	1.890	1.905	2.110	2.120	2.205	2.255	2.205	2.135	2.085	2.085	2.145	2.245	2.290	2.240	2.160	2.095	2.050	2.045	2.080	2.140	2.160	2.140	2.050	1.950
15	1.850	1.870	1.975	2.125	2.245	2.290	2.240	2.160	2.095	2.090	2.150	2.240	2.270	2.245	2.175	2.125	2.100	2.125	2.175	2.210	2.215	2.180	2.050	1.910

Figure 194

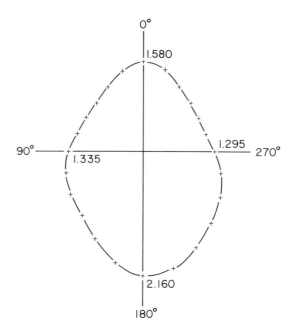

Figure 195

After casting, the preform may be shaped by cutting or grinding. It is necessary to mount the preform on a dividing head or indexing table on the numerically controlled mill, since it must be rotated by 15-deg increments to be properly shaped. A vertical-spindle positioning mill is suitable, the machine being used to mill the side of the preform in longitudinal cuts. If there is Z-axis control, then either the side or the top of the preform may be milled, surface finish deciding which method to use.

Data for the control tape may be hand calculated, since generous tolerances are allowable. If APT language is to be used, then each 15-deg profile requires a TABCYL definition. Some of the profile points may require adjustment to suit the computer approximation.

Figure 196 shows the profiles of the model limb at 60 deg and 165 deg. TABCYL definitions for these profiles were given in the n/c machine coordinates as defined in Fig. 196. The computer analyses for these two TABCYL's are given in Figs. 197 and 198.

Questions

1. Determine the cubic equations for the two TABCYL's of Fig. 197 and 198. Plot the curves between the first two points and the second two points of the TABCYL's.

Chap. 17 / Creativity in Numerical Control   337

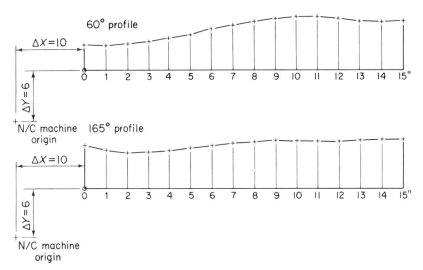

Figure 196

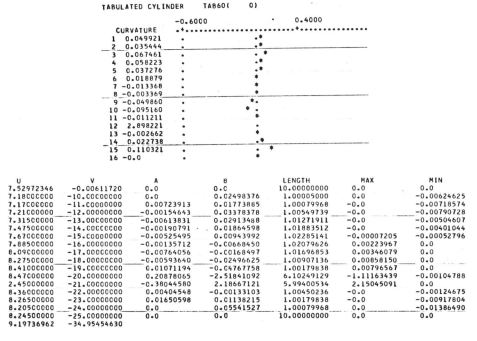

Fig. 197. TABCYL analysis for 60° profile.

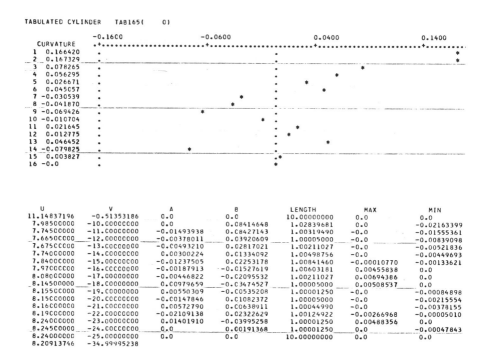

Fig. 198. TABCYL analysis for 165° profile.

2. Using small pieces of aluminum plate, design and build a fluidic tape reader to read punched tape one character at a time. Small $\tfrac{1}{16}$-in. fluidic tubing should be used to supply air to the reading manifold. Read out the hole signals as pressures on manometers, as Schmitt trigger signals, or any other method. Consider the effect of air leakage around the tape in your design.

   Try to determine the maximum reading speed of your device in characters per second.

3. Devise a fluidic logic circuit to read out the numbers 1 to 9 from a fluidic tape reader, using turbulence amplifiers and AND circuits.

4. Devise a method of determining the shear strength of paper. Find the shear strength of paper.

   What force is required to punch EIA or ASCII characters into paper tape? Assume a maximum number of holes per character of seven.

5. Investigate the procedures and economics of producing artificial limbs by numerical control and write a technical paper on your findings suitable for presentation to a technical society. Unless you feel competent

to deal with the subject, avoid the matter of data acquisition from the surviving limb.

Most artificial limbs are AK's and BK's. It is not necessary to extend the discussion to artificial arms, therefore.

Consider with some thoroughness the following topics:

a. Computer programming
probable cost
in-house or rented computer time
programming procedures
use of a MACRØ
use of standard programs
b. Machining
specification of a suitable n/c mill
suitable n/c mills on the market
number of axes
tape reading speed
contouring or positioning control
time and cost estimates for setup of preform and machining
(A cost of about $30 per hour, labor and overhead included, should be anticipated, but check this figure.)
c. Overall Cost Comparisons
present method
n/c method
d. Recommendations
Does the method offer some possibility of success?
Is it worth a research grant?
How much funding would be required for a feasibility study extending over 18 months?

ANSWERS TO QUESTIONS IN THIS BOOK

In such subjects as mathematics, we are accustomed to thinking in terms of one right answer and, of course, wrong answers. This is true of theoretical problems only. In numerically controlled operations, there are no right answers (though there are wrong answers), but only acceptable answers or solutions. An acceptable solution is any solution that produces acceptable results, and there may well be several such acceptable solutions. Therefore, only a numerically controlled machine, or, in the case of the computer programming of Part III of this book, a computer, can tell you whether you have or have not the correct solution. The problem of "correct" solutions to the questions in this book is complicated by the small differences in programming requirements between one computer and another. In numerical control, as in real life, you cannot go to someone else to pass judgment on your work; you seek out the acceptable solution yourself. You are the one working on the problem and, therefore, the only one sufficiently informed to make the decision of acceptability of results.

However, in the early stages of learning the methods of numerical control, there is some value in working out practice problems, and the student may wish to have some solutions with which to compare his work. Solutions to some of the problems in the book are presented here. These solutions may be consulted, provided the student understands when a departure from the solution given here is, or is not, acceptable.

CHAPTER 2

1. The number of holes in the DELETE character is determined by the parity and also the requirement that the erroneous character must be obliterated.
2. An absence of any hole pattern is an absence of any information. Zero is an item of information; it is not nothing.
3. M00 is an optional stop.
   M02 is a terminal stop.
   M06 is a stop for tool change.
6. $EOB = 2^7 = 128$
   $TAB = 2^5 + 2^4 + 2^3 + 2^2 + 2^1 = 62$
   $DELETE = 2^6 + 2^5 + 2^4 + 2^3 + 2^2 + 2^1 + 2^0 = 127$
   $5 = 2^5 + 2^2 + 2^0 = 37$
   $9 = 2^4 + 2^3 + 2^0 = 25$
8. Diameter of drum 0.76394 in., 2.4 in. circumference, and 0.0036 in. tolerance allowed on circumference.

CHAPTER 3

1. H010 G81 X12650 Y06625 M51
   12 TAB's
   H011 G81 X16125 Y09750
   (The program should begin with an EOB.)
2. (a) H006 G79 X06600 Y03500 M51
       N007   Y07500
   (b) H006 G78 X06600 Y03500 M51
       (CYCLE START button)
       N007 G79   Y07500
       N008 G78   (unclamp)
       N009 G81
3. H001 G81 X10437 Y05625 M51   bolt hole
   N002   X12313   bolt hole
   N003   Y03500   bolt hole
   N004   X10437   M06   bolt hole
4. H005   X11375 Y06280 M52   dowel hole
   N006   Y03470 M06   dowel hole
   H007   Y05334 M53   dowel hole
   N008   Y04416 M06   dowel hole
   N009   X11063 Y04875 M54   dowel hole
   N010   X11687   M02   dowel hole

342 *Answers to Questions in this Book*

5. H001 G81 X10500 Y14000 M51   corner hole $\frac{3}{8}$
   N002   Y03500   corner hole $\frac{3}{8}$
   N003   X19500   corner hole $\frac{3}{8}$
   N004   Y14000 M06   corner hole $\frac{3}{8}$
   H005   X15000 Y05500 M52 $\frac{27}{64}$ tap drill
   H006 G80   M06   tool change only
   N007 G81   Y12000 M53 $\frac{5}{16}$ drill, bolt circle
   N008   X16664 Y10791 $\frac{5}{16}$ drill, bolt circle
   N009   X16029 Y08834 $\frac{5}{16}$ drill, bolt circle
   N010   X13971 $\frac{5}{16}$ drill, bolt circle
   N011   X13336 Y10791 M06 $\frac{5}{16}$ drill, bolt circle
   H012   X15000 Y10250 M02 $\frac{5}{16}$ drill, bolt circle

6. H001 G80 X15000 Y05500 M50
   N002 G78   M51
   N003 G79   Y05000
   N004   X11500
   N005   Y06000
   N006   X18500
   N007   Y05000
   N008   X14900
   N009 G78   unclamp
   N010 G81   M02

7. The two holes #23 drilled through may require a woodpecker drilling sequence and might, therefore, be done manually on an n/c machine without a woodpecker drilling canned cycle.
   H001 G81 X10396 Y10868 M51   #15 drill
   N002   Y11133 M06   #15 drill
   (countersinking not included here)
   N005   X12418 Y11350 M53 $\frac{3}{32}$ drill
   N006   Y10650
   N007   X12801 Y11000 M06
   (countersinking not included here)
   H010   X10446 Y10625 M55   holes C
   N001   Y11375
   N012   X12670 Y11300
   N013   Y10700 M06

CHAPTER 4

5. $S/2 = \sqrt{R^2 - (R-h)^2}$
   $S = 2\sqrt{2Rh - h^2}$
   But since h is very small,
   $S = 2\sqrt{2Rh}$

Sec. 7.2

1. Radius of the tool path is 3.000 in.
$$\theta = 2\cos^{-1}\frac{2.998}{3.002} \quad \text{and} \quad \theta = 6 \text{ deg}$$
Use half-tangents of 3 deg for first and last increments.
The points on the tool path from A to F are these:

Point	Angle (deg)	X	Y
A	30	10500	12598
1	33	116776	12487
2	39	11887	12331
F	42	12046	12229

2. $\cos 2°35' = 0.99898$
   $\cos 3°37' = 0.99803$
   Angular range is $1°02'$.

Sec. 7.6

In the following solutions, points on the work piece are designated P1, P2, etc. and points on the tool path are designated P1T, P2T, etc.

1.

WORK PIECE			TOOL PATH		
Point	X	Y	Point	X	Y
P1	02000	01000	P1T	01750	00750
P2	08000	01000	P2T	08000	00750
P3	09000	02000	P3T	09250	02000
P4	09000	04125	P4T	09250	04375
			P4T1	09000	04375

P5	07625	05500	P5T	07875	05500
			P5T1	07875	05750
P6	04598	05500	P6T	04531	05750
P7	02000	04000	P7T	01750	04144
P1	02000	01000	P1T	01750	00750

2.

	WORK PIECE			TOOL PATH	
Point	X	Y	Point	X	Y
P1	02000	01000	P1T	01750	00750
P2	06000	01000	P2T	06250	00750
P3	06000	01804	P3T	06250	01737
P4	07000	03536	P4T	07289	03536
P5	05000	07000	P5T	05054	07407
P6	02000	04000	P6T	01750	041035
P1	02000	01000	P1T	01750	00750

Sec. 9.7

1. A minus sign must be used with I and J dimensions where required. If the center of the circle is in the minus direction with respect to the starting point of the cut, the minus sign applies. The following program begins at the top point of the O-ring groove and travels counterclockwise. The program must begin with EOB.

```
 0  RWS
 1  T5375   T-5005  T       T       T55
 2  T-0555  T-0320  T       T-0641  T52
 3  T-0215  T-0800  T1385   T-0800
 4  T0215   T-0800  T1600
 5  T0555   T-0320  T0555   T0320
 6  T0555   T0320   T       T0641
 7  T0215   T0800   T-1385  T0800
 8  T-0215  T0800   T-1600  T
 9  T-0555  T0320   T-0555  T-0321
10  T-5375  T5005   T       T       T02535
```

CHAPTER 9

2. Yes. In moving from 0,0 to 4,6, the tool positions at 45 deg to both axes until the departure of 4 in. in X is completed, then completes the departure in Y.

3. Each step requires 0.06 second machining time. Twelve characters are read per step, or 12/0.06 sec. Reading speed should exceed 200 characters/sec.
4. There are six types of holes, labelled A to F. There are seven other hole shapes, for a total of 13 shapes and sizes of holes. Assume 13 tool changes, 13 returns to parking position, 13 moves from parking position to panel, and 13 tool-changing times of 15 seconds each.

   There are seven holes labelled A. To punch these requires one move from parking position, six moves between holes, and one move to parking position for a tool change. And so on.

CHAPTER 10

2. (a) 51
   (b) 89
   (c) 282
3. Feed rate 6 ipm
   Spindle on at 600 rpm clockwise
7. G81

APT PRACTICE QUESTIONS, CHAPTER 11

1. L1 = LINE/P1,LEFT,TANTO,C1
   L2 = LINE/P1,RIGHT,TANTO,C1
   L3 = LINE/PARLEL,L1,XLARGE,L3
2. L1 = LINE/RIGHT,TANTO,C1,RIGHT,TANTO,C2
   or L1 = LINE/LEFT,TANTO,C2,LEFT,TANTO,C1
   L2 = LINE/LEFT,TANTO,C1,RIGHT,TANTO,C2
3. C2 = CIRCLE/CENTER,P1,RADIUS,2.5
   C2 = CIRCLE/CENTER,P1,TANTO,L4
   C2 = CIRCLE/CENTER,P1,TANTO,L3
       etc.
4. C1 = CIRCLE/XLARGE,L100,YLARGE,OUT,C100,RADIUS,0.5
5. CIRC2 = CIRCLE/YLARGE,IN,CIRC3,YLARGE,IN,CIRC4,
           RADIUS,.641
   CIRC1 = CIRCLE/YSMALL,IN,CIRC3,YSMALL,IN,CIRC4,
           RADIUS,.641
6. C1 = CIRCLE/1.5,3.5,0.75
   C2 = CIRCLE/7.5,3.5,1.5
   L1 = LINE/LEFT,TANTO,C1,LEFT,TANTO,C2
   L2 = LINE/RIGHT,TANTO,C1,RIGHT,TANTO,C2

CHAPTER 12

1. CUTTER/0
   FRØM/SETPT
   GØTØ/4.375,12.995,8
   GØDLTA/0,0,−8
   GØ/ØN,CIRC2,TØ,PTSURF,ØN,PERPLN
   DRAFT/ØN
   GØLFT/CIRC2,TANTØ,CIRC4
       etc.
   GØFWD/CIRC2,ØN,PERPLN
   DRAFT/ØFF
   GØDLTA/0,0,8
   GØTØ/SETPT
   FINI

3.

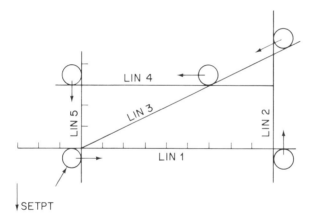

4. The following program assumes an origin at the lower left-hand corner of the part.
   PARTNØ PØCKET
   CUTTER/1
   NØPØST
   TØLER/...
   CLPRNT
   SETPT = PØINT/−6,12
   L1   = LINE/1,0,1,10
   L2   = LINE/0,4,10,4
   L3   = LINE/PARLEL,L1,XLARGE,8
   L4   = LINE/PARLEL,L2,YLARGE,6.5
   FRØM/SETPT

```
    GØ/PAST,L4,PAST,L1
    GØRGT/L1,TØ,L2
    GØLFT/L2,TØ,L3
    GØLFT/L3,TØ,L4
    GØLFT/L4,TØ,L1
    GØTØ/SETPT
    FINI
6.  PARTNØ PCHPLT
    MACHIN/FIRBR,1
    CLPRNT
    INTØL/.003
    ØUTTØL/.006
    CUTTER/.19
    SETPT  = PØINT/4,13,.75
    P1     = PØINT/4,0,.75
    P2     = PØINT/11,0,.75
    PARPLN = PLANE/SETPT,P1,P2
    P3     = PØINT/11,10,.75
    P4     = PØINT/4,10,.75
    P5     = PØINT/11.5,11.5,.75
    P6     = PØINT/0,11.5,.75
    L1     = LINE/P5,P6
    L2     = LINE/P3,P2
    L3     = LINE/P1,P2
    L4     = LINE/P1,P4
    C1     = CIRCLE/CENTER,4,10.5,.75,RADIUS,1.375
    C2     = CIRCLE/CENTER,11,10.5,.75,RADIUS,1.375
    C3     = CIRCLE/YLARGE,L3,XSMALL,L2,RADIUS,.375
    C4     = CIRCLE/XLARGE,L4,YLARGE,L3
    FLAME/ØN
    FRØM/SETPT
    GØ/TØ,C1,TØ,PARPLN,TØ,L1
    PSIS/PARPLN
    TLLFT,GØLFT/L1,TØ,C2
    GØLFT/C2,TØ,2,INTØF,L2
    GØLFT/L2,TANTØ,C3
    GØFWD/C3,TANTØ,L3
    GØFWD/L3,TANTØ,C4
    GØFWD/C4,TANTØ,L4
    GØFWD/L4,TØ,C1
    GØLFT/C1,TØ,2,INTØF,L1
    FLAME/ØFF
    GØTØ/SETPT
    FINI
```

7. Geometric statements:
   C1 = CIRCLE/0,0,0,0.75
   C2 = CIRCLE/CENTER,−1.5,0,0,RADIUS,.75
   C3 = CIRCLE/0,0,0,1.25
   SETPT = PØINT/−6,10,8
   P2   = PØINT/−2.25,0,0
   P3   = PØINT/−2.25,2.25,0
   L1   = LINE/P2,P3
   L2   = LINE/PARLEL,L1,XLARGE,7.5
   L3   = LINE/LEFT,TANTØ,C2,LEFT,TANTØ,C3
   P4   = PØINT/INTØF,L3,L2
   C4   = CIRCLE/CENTER,P4,RADIUS,0.5
   L6   = LINE/RIGHT,TANTØ,C2,RIGHT,TANTØ,C3
   P5   = PØINT/INTØF,L6,L2
   C5   = CIRCLE/CENTER,P5,RADIUS,0.5
   L4   = LINE/PARLEL,L3,YLARGE,0.5
   L5   = LINE/PARLEL,L3,YSMALL,0.5
   L7   = LINE/PARLEL,L6,YLARGE,0.5
   L8   = LINE/PARLEL,L6,YSMALL,0.5
   PARPLN = PLANE/P2,P3,P4
   Motion Statements:
   FRØM/SETPT
   GØTØ/2,1,8
   GØDLTA/0,0,−8
   GØ/TØ,L4,TØ,PARPLN,TØ,L1
   GØRGT/L1,PAST,L8
   GØLFT/L8,TANTØ,C5
   GØFWD/C5,TANTØ,L7
      etc.

8. The geometry and motion commands are similar to those of Question 7. However, C4 is thus defined:
   C4 = CIRCLE/XLARGE,OUT,C2,OUT,C5,RADIUS,2.5
   and L1 = LINE/LEFT,TANTØ,C1,LEFT,TANTØ,C2
   All motion commands are GØFWD/.

9. This is an excellent question for the beginner in APT to practice with. He should remove any apparent errors in his program, but should not assume his program to be either acceptable or wrong unless he puts it through a computer. No solution is offered here. This is not an easy program to write. The best way to solve it is to write the best possible program, run it through the computer for an error printout, and rectify the program in the light of the computer's remarks.

   The ledge to be milled contains two ramped planes. Presumably the programmer will use the vertical sides of the ledge as his drive and check surfaces. The programmer must decide whether to execute the operation clockwise or counterclockwise. When coming on to the ramped plane P6P5P2 from the right or clockwise direction, a TLOFPS command will be necessary; otherwise, the end mill will gouge the plane P4P5P2. Another difficulty arises in stopping the motion at the line P5P2.

11. The four circles are defined to three-decimal accuracy. They are not necessarily tangent to four-decimal accuracy. The definitions of any two circles must be suitably altered.

Sec. 15.13

1. Top horizontal row of holes:
   DAA = 35.75,0,0
   PAT1 = PATERN/DAA,SX(−2.375)SY(0)EX(−19.375)NH(6)
   The following defines all holes except three in the top five horizontal rows of tube holes:
   DAA = 35.75,0,0
   PAT2 = PATERN/DAA(−2.375,0)(−4.3125,−3.375)(−2.375,$
   −6.750)(−4.3125,−10.125)(−2.375,−13.5)
   PAT3 = PATERN/DAA,PAT1,SX(−2.375)SY(0)EX(−19.375)$
   EY(0)NH(6)
   Lowest horizontal row of five holes:
   DAB = 16,0,0
   PAT4 = PATERN/DAB,SX(−14.375)SY(−0.5)EX
   (−18.250)NH(5)

2. Spot drilling of PAT4:
   SPDRL,101/DP(0.1)/DAA,PAT4
   Spot drilling of PAT1 and PAT4:
   SPDRL,101/DP(0.1)/DAA,PAT1,DAB,PAT4

Sec. 15.15

   TØØL/SPDRL 0646 0.100 118.0 4.020 4.014 1200 2.0 ØN M
   TØØL/DRILL 0749 0.281 118.0 4.600 4.517 520 2.5 ØN M
   CL = 0.2
   DAA = 4.0,6.0
   BOLTCK = PATERN/AT(1.625,2.125)RADIUS(0.75)SA$
            (5.0)IA(45.0)NH(8)
   PAT = PATERN/SX(1.25)SY(0.375)DX(0.25)NH(4)
   PATS = SPDRL,0646/DI(0.1)/DAA,BOLTCK,PAT(0,0)$
          ATANGL(90)PAT, PAT(3.25,0)ATANGL(90)
   DRILL,0749/DP(0.5)/PATS

INDEX

G FUNCTION INDEX

G00, 117
G01, 100, 117, 146
G02, 100, 117, 146
G03, 100, 117, 146
G04, 117, 147
G33, 117, 147, 157
G34, 117, 157

G35, 117, 157
G78, 51
G79, 51
G80, 51
G81, 51
G84, 52
G85, 52

M FUNCTION INDEX

M00, 47, 80, 148
M01, 80, 148
M02, 47, 81, 148
M03, 81, 148
M04, 81, 148
M05, 81, 148

M06, 47, 81
M07, 81
M08, 81, 148
M09, 81, 148
M26, 125
M30, 148

COMPUTER VOCABULARY INDEX

AUXFUN, 256

CIRCLE, 223, 322
CLEARP, 256
CLPRNT, 211
COOLNT, 254
CUTTER, 211, 231
CYCLE, 256
CYLNDR, 226

DELAY, 257
DENSE, 263
DRAFT, 258

END, 255

FEDRAT, 254
FINI, 206, 210, 211
FLAME, 258
FROM, 232

GCONIC, 305
GOBACK, 238
GODLTA, 232
GODOWN, 238
GOFWD, 238
GOLFT, 235, 237
GORGT, 237
GOTO, 232
GO/TO, 236
GOUP, 238

INDIRP, 239
INDIRV, 239
INERTA, 262
INTOF, 241
INTOL, 212
ITEM, 262

LEADER, 256
LEFT, 219

LINE, 219
LOADTL, 257

MACHIN, 212, 253
MACRO, 263
MCHTOL, 212, 255

NOPOST, 212

ON, 237
ORIGIN, 254
OUTTOL, 212

PARTNO, 211
PAST, 237
PATERN, in APT, 268
 in AUTOSPOT, 287
PENDWN, 258
PENUP, 258
PLANE, 223
PLUNGE, 258
POCKET, 265
POINT, 219
PREFUN, 257
PRINT/3,ALL, 243
PSIS, 240

QADRIC, 304

RAPID, 257
REMARK, 213
RETRCT, 256
RIGHT, 219
ROTABL, 257

SELCTL, 257
SEQNO, 255
SPINDL, 253
STOP, 255

TABCYL, 321

TANTO, 237
TERMAC, 264
THICK, 242
TLLFT, 240
TLOFPS, 240
TLONPS, 240
TLRGT, 240
TO, 237
TOLER, 244

TRANS, 254

VECTOR, 227

XLARGE, 219, 255

YLARGE, 219, 255

ZSURF, 233

SUBJECT INDEX

Absolute Positioning, 11
ADAPT, 273
Adaptive Control, 183
Address Codes, Table of, 116
Angle Cuts, 59
APT, 199
Arc Center Offsets, 99, 155
ASCII Tape Code, 27
Automatic Tool Changer, 77
AUTOSPOT, 281
Axis, 10
   Rotary, 13

Binary Coded Decimal, 29
Binary Number System, 29
Bit, 31
Block of Information, 31
   for Cintimatic Mill, 45, 127
   for Gorton Tapemaster, 120
   for Lathes, 137
   for Milwaukee-Matic Model II, 79
   for Moog, 69
   for Slo-Syn, 171
Buffer Storage, 15, 167

Cards, Punched, 20

Check Surface, 234
Chords, 94
Cintimatic Mill, 42, 124, 184
   Turret Drill, 125, 184
Circle Contouring, 60, 94, 99
Circular Interpolation, 99, 168
Closed Loop Control, 169
Computers, 21
Contamination, of Oil, 177
Contouring Control, 14
Contouring Machines, 177
Curve Fitting, 102

DASH (AUTOSPOT), 286
Datum Points, AUTOSPOT, 286
Delete Key, 26
Depth Cams, 49
Direct Feed Rate, 155
Direct Numerical Control, 15
Drafting, 181
Drill Point Length, 48
Drive Surface, 234
Dwell, 147
   with Delete Code, 52

EIA Tape Code, 28
Electronic Data Processing, 4

Index

Feed Rate Number, 120, 148
Fillets, **AUTOSPOT**, 299
Fixed Automation, 4
Fixed Block Format, 35, 67
Flame Cutting, 182
Flexowriter, 25
Formats, 31
   word address, 31
   tab sequential, 35
   fixed block, 35
Froth-Pak, 337

Gage Height, 125
Geometric Names, 209
Geometric Statements, **APT**, 213
G functions, 50
Gorton Tapemaster, 76, 119, 184

Horizontal Spindle Mill, 76
Hydraulic Oil, 177

Incremental Positioning, 11
Information Block, 31
   for Cintimatic Mill, 45, 127
   for Cintimatic Turret Drill. 127
   for Gorton Tapemaster, 119
   for Lathes, 137
   for Milwaukee-Matic Model II, 79
   for Moog, 69
   for Slo-Syn, 53, 171
Inspection, 180

Lathes, 135
Lead, of Thread, 157
Linear Interpolation, 94, 168. 206

Machining Center, 77

Machining Statements, **AUTOSPOT**, 289
Magic Three, 120
Magnetic Tape, 19
Milling, **AUTOSPOT**, 297
Milwaukee-Matics, 77, 184
Miscellaneous Functions, 36, 47
Moment of Inertia, 261
Monarch turN/Center, 142
Moog Machines, 68, 184
Multiple Intersections, **APT**, 241

Nested Definitions, 244
Numerical Control, 4
   Procedures, 8

Oil Filters, 178
Open Loop Control, 169

Parabolic Interpolation, 101, 168
Parity Check, 29
Part Surface, 234
Patterns, **APT**, **AUTOSPOT**, 268, 287, 292
Perishable Tooling, 179
Picture Frame Fixture, 178
Pocket Milling, 57, 265, 299
Positioning Control, 14
Postprocessor, 9, 205, 251, 294
Preparatory Functions, 50, 117
Process Control, 4
Processor, for **APT**, 215
Programming—
   of Cintimatic Mill, 45, 124
   of Cintimatic Turret Drill, 124
   of Gorton Tapemaster, 119
   of Milwaukee-Matic Model II, 78
   of Monarch turN/Center, 142
   of Moog 83-1000, 69
   of Slo-Syn, 171

Punched Cards, 20
Punch Press, 182

Quadric Function, 106
  in APT, 304
Quickpoint, 275

Register, 23
Rounding-Off Errors, 62

Secants, 94, 206
Sequence Number, 46
Servomotor, 169
Set-Up Point, 44
Sign Convention, for Increments, 53, 118
  for Lathes, 135
Slo-Syn, 53, 171
  -Motor, 170
Software, 6
Specification Statements, AUTOSPOT, 285
Start-Up Statements, 236
Subplate, 45

Tab Sequential Format, 35, 53
Tangents, 94

Tape, Paper, 24
  Codes, 27
  Formats, 31
Taper Cuts, 141
Tape Reader, 14, 165
Tape Reading Speed, 115
Tape Search, 47
Thread Cutting, 156
Tool Code, Milwaukee-Matic, 80
Tool End, APT, 231
Tooling, 178
Tool Movement Control, APT, 234
Tool Offset, 60, 106
  in Lathes, 138
  in AUTOSPOT, 299
Tool Offset Calculations, 106
Tool Statements, AUTOSPOT, 295

UNIAPT, 274

Welding, 182
Word, 31
Word Address Format, 31

Z-Axis, 11
Zero Offset, 45
Zero Point, 45

SAINT JOSEPH'S COLLEGE, INDIANA
TJ1189 .P36   ISJA
Patton / Numerical control: practice and application

3 2302 00038 7292